TOMCAT'S BIG
CB HANDBOOK

CITIZEN BAND COMMUNICATIONS

TOMCAT'S BIG
CB HANDBOOK

by Tom Kneitel
"Tomcat"/SSB-13

CRB Research Books, Inc.

P. O. Box 56
COMMACK, NEW YORK 11725

Dedicated to Judy (SSB-99), to
Walter, Danny, Noelle, and Tommy II.

Third Printing, Revised September, 1992.

Artists' credits:

John Kane: Pages 9, 49
Ken Kneitel: Page 216
Harold Perry: Pages 16, 23, 79, 80, 83, 117, 119, 126, 134, 141, 156, 165, 170, 172, 173, 176, 192, 196, 199, 214
Henry Projansky, W1GSI: Front cover, 123, 129, 149, 197, 198

Cover, layout and book design by the CRB Research Books, Inc. Graphics Department.

ISBN: 0-939780-18-6

TABLE OF CONTENTS

(continued next page)

There have been scores of CB books, both technical and otherwise, over the years. Although many have been excellent, none have filled a gap I've always noticed in the information contained in CB books. In 1976, I was approached by a major publisher with the idea of bringing out a CB book. I said I wanted to do a no-holds-barred book that spoke to operators in their own language, told them things that nobody else had ever included in a CB book. They agreed, but hoped that I would still remember that there were some areas that would have to be handled "delicately." When they saw my manuscript they freaked, ultimately gentrifying it into a mild-mannered, but highly popular and well-selling book that wasn't at all what I had hoped it would be.

In the years that have passed since 1976, CB somewhat faded in popularity, but eventually rebounded with as much vigor and potential as ever-- it seems even better now than it was in the 1970's!

This seemed like the time to bring out the type of book that they said couldn't be written-- a frank, blunt, and candid look at contemporary CB, including a look into its darker corners and more flamboyant and outrageous aspects, customs, traditions, foibles, and denizens, legal and illegal. I think (and hope) it will bring the new and old 27 MHz operator an honest, comprehensive, and wide ranging guide to successful and exciting participation in the dynamic world of 27 MHz communications-- as well as an understanding of the gray areas that have too long been relegated to the shadows. Iconoclast that I am, I have perhaps also managed to bash in some of the more senseless and irritating CB traditions that are near and dear to the hearts of many. Hope not too many of your favorites have been too badly smashed in.

I should like to point out that many of the very clever cartoon illustrations in this book were drawn by Harold Perry, a brilliant and zany illustrator who originally produced these works for me when I was the Editor of S9 Magazine. Permit me to thank Judy, SSB-99 (ex-2W8830 and later KBDA3756), for urging me to finally write this book after talking about it and planning it in my head for so many years.

Tom Kneitel
"Tomcat"/SSB-13
K2AES

1

CB: The Long Road From Then To Now

During the closing months of World War II, the Federal Communications Commission (FCC) conducted a series of fact-finding inquiries and hearings that cluminated in a "Report of Proposed Allocations" dated 16 January, 1945, covering the frequency range from 25 MHz to 30 GHz, and bearing the designation "Docket No. 6651." In that report, the FCC proposed a peacetime allocation of the frequency spectrum above 25 MHz, to provide for the reasonably known or estimated needs of all segments of the American public whether local governments, industries, or private individuals. To meet the latter need-- to make possible the fullest practicable development of private radio communications within the limits set by other demands for assignments in the spectrum-- the FCC proposed to allocate the band 460 to 470 MHz to a new radio service to be called the Citizens Radio Service.

The Citizens Radio Service, at that time, was envisioned and defined by the FCC as a radiocommunication service of fixed, land, and mobile stations. The stated purpose was to provide for the use of radio facilities by any citizen of the United States for private short-distance radio communications, radio signalling, or the control of remote objects or devices by means of radio. As opposed to the Amateur Radio Service, which had been established many years earlier to accomodate privately owned stations operated by radio hobbyists and experimenters, all of whom held licenses obtained after passing a government-administered technical, operating, and code (CW) proficiency test, the Citizens Radio Service was to be different. The concept was to get non-technical people on the air with licenses that would be issued upon completion of a simple application form-- and with the understanding that their radio equipment would be used for utilitarian (i.e. substantive, non-hobby) communications, primarily between different units of the same licensee.

Such a service wasn't possible before WWII because sufficient frequencies in the higher portions of the radio spectrum had not been developed to meet the needs of the more pressing services. That was the situation despite the fact that the usable radio spectrum was already a hundred times larger than in 1925, when for all practical purposes it ended around 3 MHz. By 1940, the ceiling of formal FCC allocation was 300 MHz, but only the frequencies up to possibly 100 MHz were really in effective use.

Because the usable frequency spectrum was rising like a thermometer, the FCC made provisions for the proposed CB service as far down the spectrum as possible to make it more readily available for use. However, owing to such needs as those of FM/TV broadcasters, public safety and government agencies, aviation, the lowest slot that could be found for CB was the 460-460 MHz band. This spot was considered well suited for CB for two reasons. First, no "skip" (distant station) interference would be experienced, and, second, it was felt that because of their experience in making radio equipment for the military services, manufacturers would be able to design and construct sets for these frequencies soon after the end of the war.

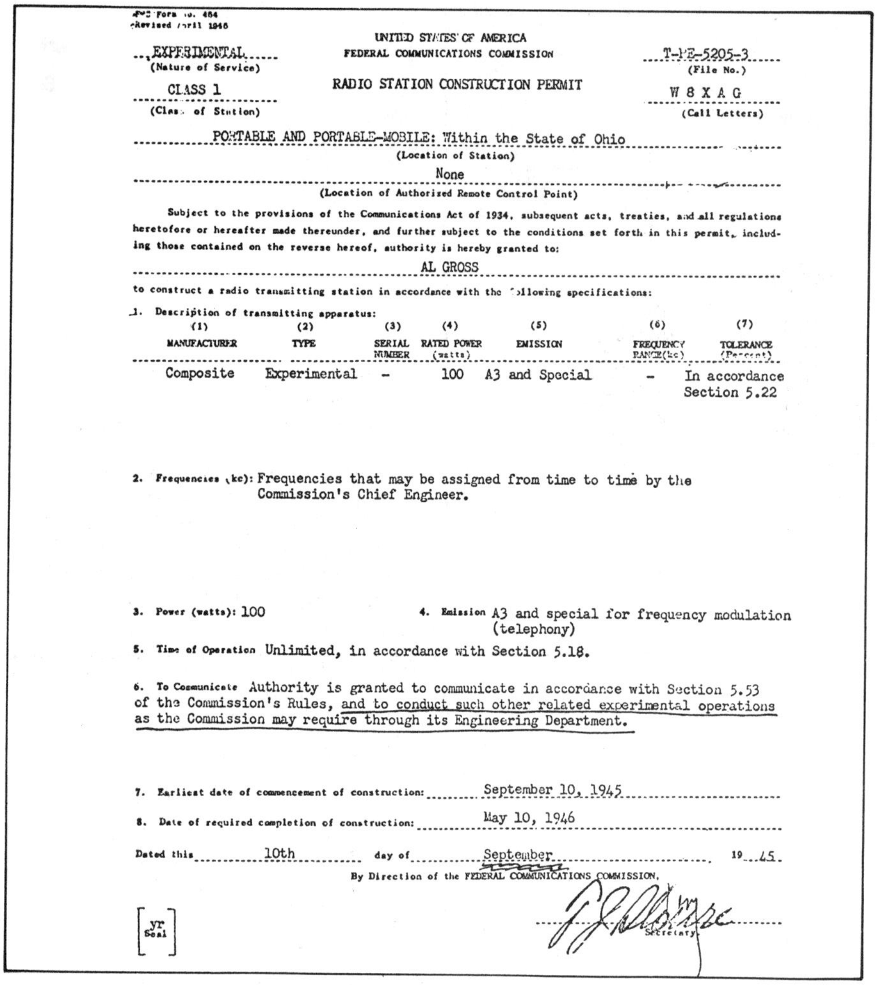

FCC Form No. 484
Revised April 1946

UNITED STATES OF AMERICA
FEDERAL COMMUNICATIONS COMMISSION
RADIO STATION CONSTRUCTION PERMIT

EXPERIMENTAL (Nature of Service)
CLASS 1 (Class of Station)
T-PE-5205-3 (File No.)
W 8 X A G (Call Letters)

PORTABLE AND PORTABLE-MOBILE: Within the State of Ohio
(Location of Station)

None
(Location of Authorized Remote Control Point)

Subject to the provisions of the Communications Act of 1934, subsequent acts, treaties, and all regulations heretofore or hereafter made thereunder, and further subject to the conditions set forth in this permit, including those contained on the reverse hereof, authority is hereby granted to:

AL GROSS

to construct a radio transmitting station in accordance with the following specifications:

1. Description of transmitting apparatus:

(1) MANUFACTURER	(2) TYPE	(3) SERIAL NUMBER	(4) RATED POWER (watts)	(5) EMISSION	(6) FREQUENCY RANGE (kc)	(7) TOLERANCE (Percent)
Composite	Experimental	-	100	A3 and Special	-	In accordance Section 5.22

2. Frequencies (kc): Frequencies that may be assigned from time to time by the Commission's Chief Engineer.

3. Power (watts): 100

4. Emission A3 and special for frequency modulation (telephony)

5. Time of Operation Unlimited, in accordance with Section 5.18.

6. To Communicate Authority is granted to communicate in accordance with Section 5.53 of the Commission's Rules, and to conduct such other related experimental operations as the Commission may require through its Engineering Department.

7. Earliest date of commencement of construction: September 10, 1945

8. Date of required completion of construction: May 10, 1946

Dated this 10th day of September 19 45.

By Direction of the FEDERAL COMMUNICATIONS COMMISSION.

Secretary

Seal

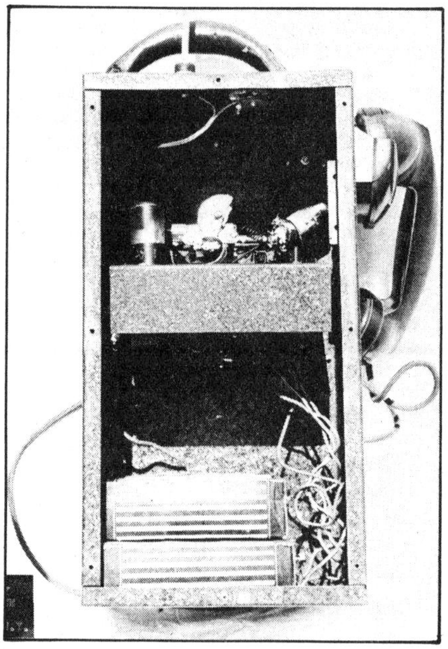

At the left is a copy of the first 460 MHz CB-band license, Experimental station W8XAG. It was issued to Al Gross of Cleveland, OH. At the right, John Mulligan's first 465 MHz transceiver wasn't much to look at, but it was nonetheless historic.

On The Air...At Last!

The first developmental (then called Class 2 Experimental) authorization for the operation of a station in the CB service was issued to Mr. John M. Mulligan, a radio engineer of Elmira, NY, on 14 February, 1947. It was followed in June of that year by proposed technical standards for equipment to be used in the service, which were slightly modified and adopted by the FCC to become effective on 1 December, 1947. These standards stipulated the use of type-approved equipment to be used in either of two classes of stations to be operated in the Citizens Radio Service.

The first, called "Class A," was to be subject to fairly strict technical standards, roughly similar to those imposed upon other services using comparable frequencies, and would permit up to 50 watts (input) when operating on frequencies in the 460-462 or 468-470 MHz portions of the band, but only 10 watts (input) elsewhere.

The second, "Class B," stations were limited to 10 watts (input) but the technical standards were such that nearly any equipment could be used so long as all operation (including tolerance and the occupied bandwidth) was confined within ±0.4% of 465 MHz. (In 1950 this was changed to ±0.5%).

The first CB equipment to be type-approved was for Class A operation.

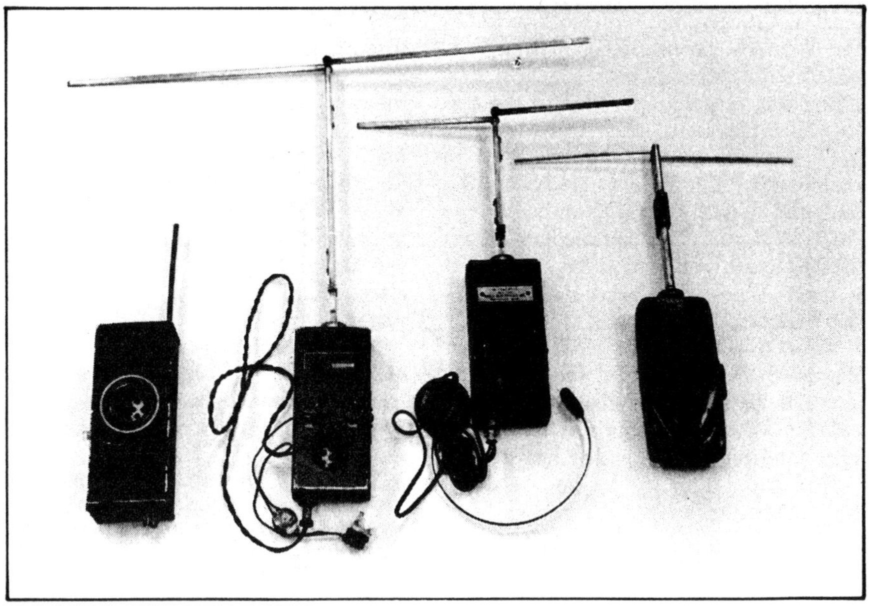

An assortment of handheld UHF transceivers built by Al Gross. At the far left is a mini-transceiver from the days when such rigs were normally contained in suitcases. Second from left is a WWII OSS (CIA) rig. Next, his first CB design. At the far right, the CB rig the FCC approved for sale to the public in early 1948.

FCC Form 505-2
United States of America
Federal Communications Commission
Not Transferable

CITIZENS RADIO STATION LICENSE

Licensee and P. O. Address (Print): THOMAS S. KNEITEL
175 WEST 93 STREET
(Number) (Street)
N.Y. 25 N.Y.
(City) (Zone) (State)
CANCELLED

Class of station	Number of transmitters	FCC type approval No.
B	1	CR 402

Not to be filled in by applicant 16—58488-2 GPO
Call Sign 2A0305
Issue date 3/9/53
Expiration date 3/9/58
Secretary.

The first CB license I held bore the callsign 2A0305, and was dated March 9, 1953. The callsign indicated that I was the 305th CB station applicant to receive a CB license issued by the FCC's New York/New Jersey area office. By then, 460-470 MHz CB was five years old. That means that only 60 licenses per year were being issued by the FCC Field Office covering all of New York and New Jersey. It's an indication of the lack of interest in UHF CB.

It remained, however, for type-approved "B" equipment to become more generally available, within a low-price range attractive to the individual, before the original intent of the FCC with respect to making two-way radio communications available to the average citizen could hope to be fulfilled. FCC records show that by early 1948, Al Gross had been given the FCC's go-ahead to merchandise a Class B unit his company, Citizens Radio Corp., had developed. Montgomery Ward immediately agreed to buy 25,000 of the hand-held transceivers. On 22 March, 1948, the FCC issued Gross the very first actual Citizens Radio Service license, 19W0001.

Still, it wasn't until the FCC relaxed some of its technical standards that there was any general commercial availability of equipment. By the autumn of 1955, Class B equipment was being marketed by Gross' company, by the Vocaline Company of America, the Stewart-Warner Company, and a few others. There was no doubt about it, the whole idea was a commercial disaster of the first magnitude. The equipment was as expensive as it was tempermental and unreliable-- and the communications range was too short to make the cost and effort worthwhile. Besides, most of the general public had never even heard of the Citizens Radio Service, and those who had, couldn't have cared less.

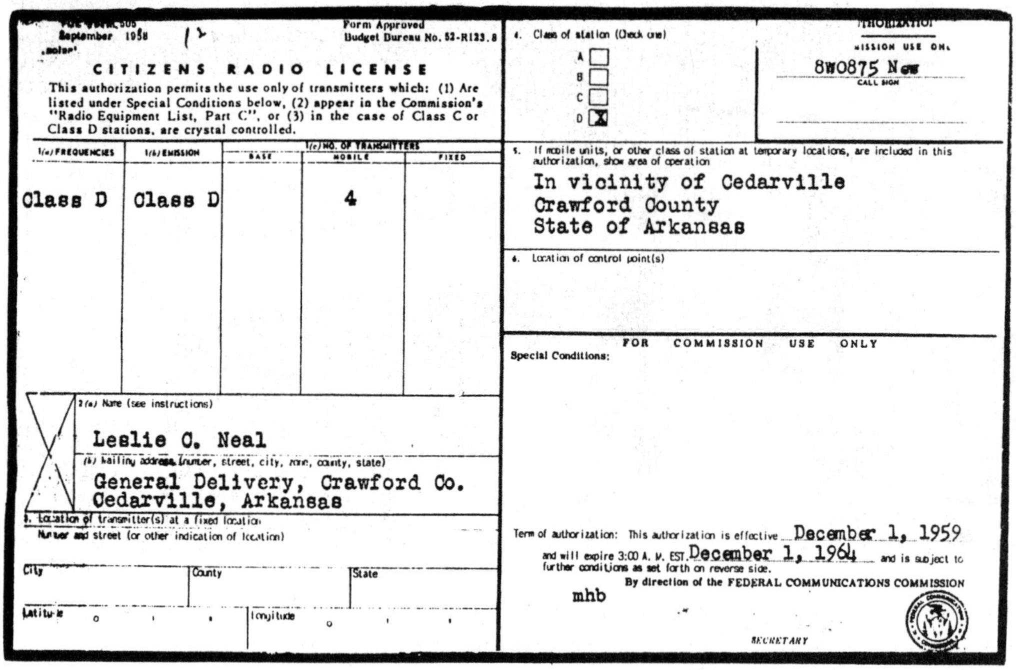

Form Approved
Budget Bureau No. 52-R133.8

CITIZENS RADIO LICENSE

This authorization permits the use only of transmitters which: (1) Are listed under Special Conditions below, (2) appear in the Commission's "Radio Equipment List, Part C", or (3) in the case of Class C or Class D stations, are crystal controlled.

1(a) FREQUENCIES	1(b) EMISSION	1(c) NO. OF TRANSMITTERS BASE	MOBILE	FIXED
Class D	Class D		4	

2(a) Name (see instructions)
Leslie C. Neal
(b) Mailing address (number, street, city, zone, county, state)
General Delivery, Crawford Co.
Cedarville, Arkansas

3. Location of transmitter(s) at a fixed location
Number and street (or other indication of location)
City / County / State
Latitude / Longitude

4. Class of station (Check one) A ☐ B ☐ C ☐ D ☒

8W0875 New
CALL SIGN

5. If mobile units, or other class of station at temporary locations, are included in this authorization, show area of operation
In vicinity of Cedarville
Crawford County
State of Arkansas

6. Location of control point(s)

FOR COMMISSION USE ONLY
Special Conditions:

Term of authorization: This authorization is effective December 1, 1959 and will expire 3:00 A. M. EST December 1, 1964 and is subject to further conditions as set forth on reverse side.
By direction of the FEDERAL COMMUNICATIONS COMMISSION
mhb
SECRETARY

A very early CB license for Class D (27 MHz) operation. This one was for station 8W0875 in Arkansas. It was issued in December of 1959 to Leslie Neal of Cedarville and allowed him to operate four units.

As soon as the FCC gave 27 MHz the go-ahead, manufacturers and dealers moved into quick action. Here's early-CB dealer Ken Krueger, 18B1652, of DeForest, WI. He specialized in Johnson Messenger rigs--they were made in Wisconsin.

CB Becomes Practical

In 1957, the FCC had to shuffle allocations in the high frequency spectrum and as a result, the CB service lost more than 6 MHz from its 460 MHz allocation. Simultaneously, the frequency band 26.96 to 27.23 MHz (then assigned to the Amateur Radio Service) was reallocated to the Citizens Radio Service (amidst considerable protest from hams), and the CB service rules were extensively revised. The "Class D" 27 MHz band was entered in the FCC's books on 3 July, 1958-- and opened for communications on 11 September, 1958.

I had tried the tricky Class B service (callsign 2A0305 issued 9 March 1953) and soon realized that it was useless. When the 27 MHz service became available, I rushed to apply for a license. After a lengthy wait for the FCC to process my license, in early 1959 it finally came through with the callsign 2W1965.

This new type of CB attracted more public interest than the earlier edition. A few stations were licensed in October and November of 1958, but had

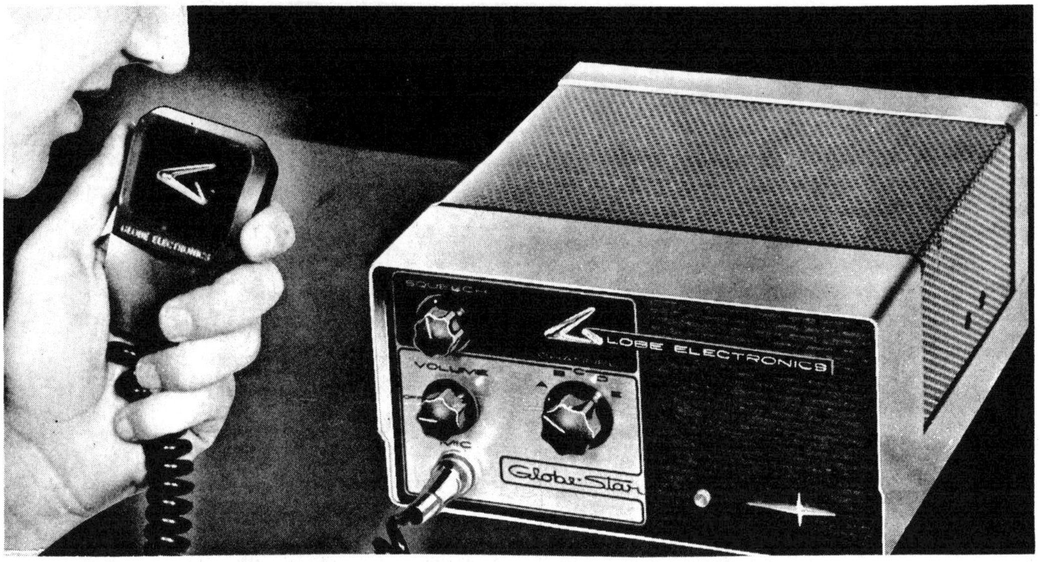

The Globe Star was a typical rig of 1962. It offered any 5 channels of your choice, a volume control, squelch, 16-tube performance, and a dual-conversion receiver. It sold for $160, including crystals for 1 channel.

reached 600 per month in January of 1959-- 5500 per month by May of 1959. Right from the start, manufacturers were producing 27 MHz equipment-- Heathkit, Multi-Products, International Crystal, Globe, Vocaline, Gonset, Eico (kit), Hallicrafters, and others. By the August, 1959, issue of Popular Electronics magazine I was writing a monthly CB column and reader mail was pouring in. Moreover, the 23 CB channels were alive with stations.

The total station count in the CB service (all classes of stations) rose from about 28,000 on 1 July 1957, to 36,000 in 1958, to 46,000 in 1959, to 126,000 in 1960, and 206,000 in 1961. As of 1 August 1961 the grand total had reached approximately 215,000 licenses-- with several units authorized for each license.

By April of 1961 it seemed logical that CB had grown to the point where it should have its own monthly magazine. After talking over the idea with several others who shared that notion, I moved from New York City to Modesto in California to turn out the first issue of a magazine called CB Horizons. In April, 1962, I left CB Horizons (which, by then had moved to Oklahoma City) and returned to New York to become the first Editor of a CB monthly called S9 Magazine, which was being brought out by the ham magazine, CQ-- I remained with S9 until 1982 when its ownership was transferred to a company that had been publishing another CB magazine for a number of years. By the time that happened, CB radio had risen (in the mid-1970's) to a major national and international mania, only to sink into almost total obscurity by the early 1980's. After several years of being relatively quiet, CB is again active, again vibrant, once again alive and buzzing away with great vigor and enthusiasm.

Of Course, Along The Way...

The creation of the 27 MHz CB service didn't quite fit the FCC's original concept for Citizens Radio. Thanks to the ingenuity of the American public, combined with the readily available equipment, plus the spectacular DX skip potentials of 27 MHz (which were especially appealing in the late 1950's and early 1960's)-- the band became a gigantic national partyline telephone, a fun place to meet and make new friends, work DX, and operate radios in a manner that seemed (to the consternation of many, including the FCC) to be somewhat of a place to operate a sort-of ham station without going through the formalities of getting a ham license.

The FCC's CB rules had never permitted such operation and the agency was immediately aghast to see its regulations so blatantly ignored on such

The early days of 27 MHz CB put CB'ers on an immediate collision course with the FCC's rules.

a massive scale. At first the FCC assumed that perhaps the regulations weren't worded clearly enough to let the public know that the 23 CB channels weren't the place for such idle chatter, that people should obtain ham licenses and leave 27 MHz available for only substantive business and personal communications. They wanted small business there, and husbands calling home from their cars to say that they'd be late for supper because of traffic jams.

Early 27 MHz CB'ers: A Wild Bunch

I'm always amused when I hear some of the more hairy-chested 27 MHz operators of today boasting of how really bad they are in comparison to the "Goody Two-Shoes" type timid souls who worshipped FCC Part 19 (the FCC's CB rules before they were renamed Part 95). The general consensus is that the massive influx of "instant CB'ers" in the mid-1970's is what pulled into CB's ranks large numbers of operators who either ignored or defied the FCC's rules-- the folks the FCC looks at with annoyance and anger.

This isn't quite true. Yes, the mid-1970's did see many new types of personalities show up, new CB lingo and buzzwords, more crowded channels, and some old operating traditions buckle under the pressure. Fact is, however, that right from day one of the 27 MHz Class D CB service the FCC was not quite pleased with the way CB was being used and realized that there were going to be problems in maintaining control over this uniquely unruly radio service.

Lots of Reasons Why

Right from the start, the original Part 19 contained so many operating restraints that the rules seemed to have been deliberately rigged so as to

Although the FCC had envisioned CB as being used by small businesses, hunters, campers, and for family members to keep in touch with one another, that wasn't the way the public perceived the new service. The public saw CB as a coast-to-coast partyline telephone-- a great hobby where you could meet new friends. Especially, it was a boon to those who spent much or all of their time at home. Some of these people spent (and still spend) the majority of their waking hours at their CB stations chatting with friends they would never have met if it wasn't for CB. This early CB'er is Rick R. Vargo, Port Huron, MI.

rule out any possible full legal use of the service. Saying much more than "hello" seemed to cause friction with several FCC regulations. For one thing, the FCC rules demanded that all CB transmissions be "substantive." Most operators had to look up the word in a dictionary to see what it meant since the FCC offered neither definition nor examples of what would comprise such a message. Operators monitored sending non-substantive messages (whatever they were in the opinion of the FCC) found themselves in hot water.

There were restrictions against off-frequency operation, against shooting skip, calling CQ, giving S-meter readings, talking too long, using certain frequencies for communicating with units of other licensees, against running overpowered transmitters, against antennas that were too high, against non use of FCC callsigns, and lots more. If you couple these restrictions with the fact that there were far fewer CB'ers on the air in the early years than what was to come later, and those operators were being enthusiastically monitored by an FCC staff that was almost identical in size even after CB grew in to the massive size it achieved in the late 1970's, you can immediately see that the probabability of getting hassled was far greater in those days than at any other time in CB's later history.

Let's face it, the way Part 19 was written, it was inevitable that there would be more than a few unintentional but unavoidable violations. On the other hand, many operators viewed the FCC's rules as being so totally unrealistic and even absurd that Part 19 became a juicy challenge to see how many sections could be ignored, trampled, defied, bent, and broken.

Overpowered rigs and illegal power boosters (linear amplifiers), transmitters with variable frequency oscillators (VFO's) were in use by many stations. Many major manufacturers were openly producing and offering the gear and every CB'er knew where it might be obtained. It seems that one the manufacturer slapped on a label reading "Illegal for use in the United States," he could sell virtually any piece of 27 MHz equipment. Companies such as Sonar, e.c.i. Courier, and others found a ready market for their 27 MHz linears. The number of ham rigs such as the Globe Scout and Johnson Thunderbolt used on 27 MHz may well have equalled the number used on the ham bands.

This equipment was very attractive to the many stations working distant stations whose signals arrived via skip from throughout the nation-- and also from other areas of the world where CB had not even been established as a radio service at that time! DX conditions were very good in those years.

CB'ers who strayed out of the band onto frequencies allocated to other services were first noted by the FCC in the mid-1960's. In addition to such stations that were licensed in the CB service, others doing this included those without any licenses as well as those holding ham licenses. In those days the FCC seemed less concerned with actually hassling such stations than they became in the 1970's-- thus giving 1970's out-of-band operators the crazy notion that they had invented and discovered the practice. In fact, most of the early days out-of-band operators came to feel that the FCC started going after such stations in response to several high-profile 1970's era organizations that were organized for purposes of encouraging stations to hop over to unauthorized frequencies outside the confines of the band.

While today there are 40 CB channels, until 1977 the CB band ended at Channel 23 (27.255 MHz). Frequencies 27.235 and 27.245 Mhz (presently CB Channels 24 and 25) were not allocated to CB in the early days. Nevertheless, they were heavily populated by CB'ers, being unofficially known as Channels 22A and 22B. Equipment would easily operate on these non-channels by the simple insertion of the proper plug-in crystals-- some CB rigs were even being manufactured with positions on their crystal switches left open for 22A and 22B. Later, when crystal synthesized rigs became popular, several sets still came factory-ready for all-channel operation, including these two bootleg frequencies.

Switcheroo

In the early days, all CB rigs required the use two plug-in crystals per channel, one for transmit and one for receive. Since the receive crystal was usually cut for a frequency 455 kHz lower than the transmitting crystal for the same frequency, it seemed only natural for someone to see what would happen if the socket positions for the transmit and receive crystals were reversed. That meant that a Channel 9 receive crystal would let the rig transmit on 26.61 kHz, well below the low frequency edge of the band. Even a Channel 23 receive crystal (26.80 MHz) was still lower than the band edge. Thus, a whole set of phantom non-channels became established outside the band limits, offering the operators alternate or "private" frequencies for DX'ing and other (sometimes nefarious) purposes.

Not Unnoticed

Not that the FCC wasn't aware of such affronts to the agency's authority. Many the operator faced the FCC's own brand of justice for skirting the rules. Not all went quietly, mind you, and there were legendary CB'ers in those days, such as Ernie Walker of New Mexico, who made an awful lot of noise before the FCC was able to finally remove his "Friendship Station" from the air-- a station that boldly and proudly worked skip stations on what seemed to many to be a 'round-the-clock basis!

Prior to the mid-1970's, most CB stations did not use CB "handles" for identifying their stations. Stations operating on unauthorized frequencies used an ever-changing array of code numbers for identification. Usually, however, regular stations used their FCC assigned callsigns-- with sideband stations also using their club numbers in addition to the FCC callsigns.

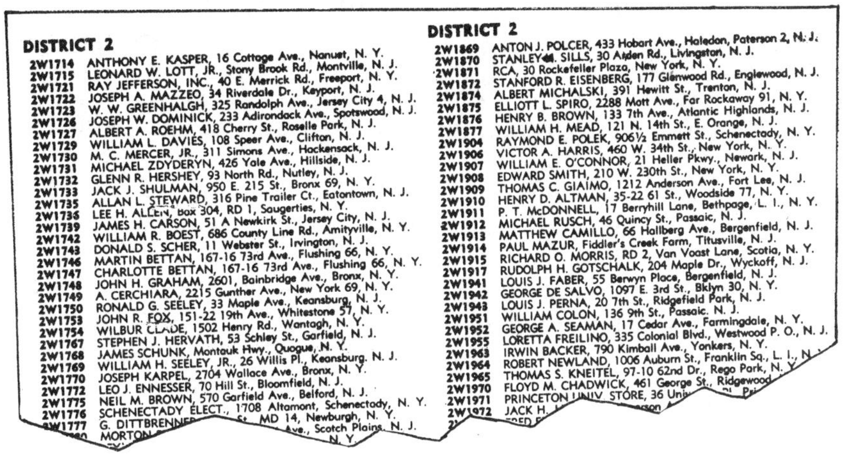

DISTRICT 2

2W1714 ANTHONY E. KASPER, 16 Cottage Ave., Nanuet, N. Y.
2W1715 LEONARD W. LOTT, JR., Stony Brook Rd., Montville, N. J.
2W1721 RAY JEFFERSON, INC., 40 E. Merrick Rd., Freeport, N. Y.
2W1722 JOSEPH A. MAZZEO, 34 Riverdale Dr., Keyport, N. J.
2W1723 W. W. GREENHALGH, 325 Randolph Ave., Jersey City 4, N. J.
2W1726 JOSEPH W. DOMINICK, 233 Adirondack Ave., Spotswood, N. J.
2W1727 ALBERT A. ROEHM, 418 Cherry St., Roselle Park, N. J.
2W1729 WILLIAM L. DAVIES, 108 Speer Ave., Clifton, N. J.
2W1730 M. C. MERCER, JR., 311 Simons Ave., Hackensack, N. J.
2W1731 MICHAEL ZDYDERYN, 426 Yale Ave., Hillside, N. J.
2W1732 GLENN R. HERSHEY, 93 North Rd., Nutley, N. J.
2W1733 JACK J. SHULMAN, 950 E. 215 St., Bronx 69, N. Y.
2W1735 ALLAN L. STEWARD, 316 Pine Trailer Ct., Eatontown, N. J.
2W1736 LEE H. ALLEN, Box 304, RD 1, Saugerties, N. Y.
2W1739 JAMES H. CARSON, 51 A Newkirk St., Jersey City, N. J.
2W1742 WILLIAM R. BOEST, 686 County Line Rd., Amityville, N. Y.
2W1743 DONALD S. SCHER, 11 Webster St., Irvington, N. J.
2W1746 MARTIN BETTAN, 167-16 73rd Ave., Flushing 66, N. Y.
2W1747 CHARLOTTE BETTAN, 167-16 73rd Ave., Flushing 66, N. Y.
2W1748 JOHN H. GRAHAM, 2601, Bainbridge Ave., Bronx, N. Y.
2W1749 A. CERCHIARA, 2215 Gunther Ave., New York 69, N. Y.
2W1750 RONALD G. SEELEY, 33 Maple Ave., Keansburg, N. J.
2W1753 JOHN R. FOX, 151-22 19th Ave., Whitestone 57, N. Y.
2W1754 WILBUR CLADE, 1502 Henry Rd., Wantagh, N. Y.
2W1767 STEPHEN J. HERVATH, 53 Schley St., Garfield, N. J.
2W1768 JAMES SCHUNK, Montauk Hwy., Quogue, N. Y.
2W1769 WILLIAM H. SEELEY, JR., 26 Willis Pl., Keansburg. N. J.
2W1770 JOSEPH KARPEL, 2704 Wallace Ave., Bronx, N. Y.
2W1772 LEO J. ENNESSER, 70 Hill St., Bloomfield, N. J.
2W1775 NEIL M. BROWN, 570 Garfield Ave., Belford, N. J.
2W1776 SCHENECTADY ELECT., 1708 Altamont, Schenectady, N. Y.
2W1777 G. DITTBRENN[illegible] MD 14, Newburgh, N. Y.
[illegible] MORTON [illegible] Ave., Scotch Plains, N. J.

DISTRICT 2

2W1869 ANTON J. POLCER, 433 Hobart Ave., Haledon, Paterson 2, N. J.
2W1870 STANLEY M. SILLS, 30 Arden Rd., Livingston, N. J.
2W1871 RCA, 30 Rockefeller Plaza, New York, N. Y.
2W1872 STANFORD R. EISENBERG, 177 Glenwood Rd., Englewood, N. J.
2W1874 ALBERT MICHALSKI, 391 Hewitt St., Trenton, N. J.
2W1875 ELLIOTT L. SPIRO, 2288 Mott Ave., Far Rockaway 91, N. Y.
2W1876 HENRY B. BROWN, 133 7th Ave., Atlantic Highlands, N. J.
2W1877 WILLIAM H. MEAD, 121 N. 14th St., E. Orange, N. J.
2W1904 RAYMOND E. POLEK, 906½ Emmett St., Schenectady, N. Y.
2W1906 VICTOR A. HARRIS, 460 W. 34th St., New York, N. Y.
2W1907 WILLIAM E. O'CONNOR, 21 Heller Pkwy., Newark, N. J.
2W1908 EDWARD SMITH, 210 W. 230th St., New York, N. Y.
2W1909 THOMAS C. GIAIMO, 1212 Anderson Ave., Fort Lee, N. J.
2W1910 HENRY D. ALTMAN, 35-22 61 St., Woodside 77, N. Y.
2W1911 P. T. McDONNELL, 17 Berryhill Lane, Bethpage, L. I., N. Y.
2W1912 MICHAEL RUSCH, 46 Quincy St., Passaic, N. J.
2W1913 MATTHEW CAMILLO, 66 Hallberg Ave., Bergenfield, N. J.
2W1914 PAUL MAZUR, Fiddler's Creek Farm, Titusville, N. J.
2W1915 RICHARD O. MORRIS, RD 2, Van Voast Lane, Scotia, N. Y.
2W1917 RUDOLPH H. GOTSCHALK, 204 Maple Dr., Wyckoff, N. J.
2W1941 LOUIS J. FABER, 55 Berwyn Place, Bergenfield, N. J.
2W1942 GEORGE DE SALVO, 1097 E. 3rd St., Bklyn 30, N. Y.
2W1943 LOUIS J. PERNA, 20 7th St., Ridgefield Park, N. J.
2W1951 WILLIAM COLON, 136 9th St., Passaic. N. J.
2W1952 GEORGE A. SEAMAN, 17 Cedar Ave., Farmingdale, N. Y.
2W1955 LORETTA FREILINO, 335 Colonial Blvd., Westwood P. O., N. J.
2W1963 IRWIN BACKER, 790 Kimball Ave., Yonkers, N. Y.
2W1964 ROBERT NEWLAND, 1006 Auburn St., Franklin Sq., L. I., N. [illegible]
2W1965 THOMAS S. KNEITEL, 97-10 62nd Dr., Rego Park, N. Y.
2W1970 FLOYD M. CHADWICK, 461 George St., Ridgewood [illegible]
2W1971 PRINCETON UNIV. STORE, 36 Uni[illegible]
2W1972 JACK H. [illegible]

At the left, the first CB callbook. It was a pamphlet that sold for $1. At the right, part of a page from the larger 1960 edition. In those days my callsign was 2W1965, listed in the right-hand column of the page shown.

Callsigns were used because, in the early days, there were CB callbooks published that listed every station with its operator's name and address. This made QSL'ing very simple. CB handles were first noticed being used in the late 1960's, probably because the callbooks were no longer being published. Also, it became obvious to operators that they stood a far less chance of getting caught by the FCC if they hid behind handles instead of announcing callsigns. Within a few years after the first use of CB handles, the use of callsigns had virtually vanished.

Rulebreakers were caught, nevertheless, and the FCC worked at this task on an unrelenting basis. At first, in the early days, the FCC's approach was more benign. The agency started out by issuing supposedly impressive sounding proclamations in order to set the record straight and explain to CB'ers that perhaps they had misunderstood the rules and just didn't realize that the FCC intended the CB service to be something vastly different than a quasi-ham radio service. Unfortunately, CB hadn't turned out as planned.

The FCC seemed to be under the naive impression that the problem was that CB'ers had somehow failed to comprehend what was expected of

CB handles began slowly in the very early days and by the mid-1960's a number of stations were using them. Still, QSL's seldom made any mention of handles. By the late 1960's, handles started to show up on QSL's. This 1967 QSL from Appy Mare and Appy Kid of Portland, OR was the first one I ever saw that was strictly for DX work--not making any reference to a regular callsign. It marked the start of an era.

them and realize that licensees simply didn't go around totally ignoring all of the agency's rules and regulations on a wholesale basis. It just wasn't done, no other licensees had ever dared try anything of that nature. The idea was to spell it out the service to its users in such clear and concise terms that there couldn't possibly be any further misunderstanding. Then everything would finally settle down and CB'ers would behave.

Clear Cut

Probably the best example of this approach was the very first of these clarification proclamations. This interesting document issued forth on 7 December 1959 when the 27 MHz band activities had been going for little more than a year. FCC Bulletin #81482 was issued in an effort to toss cold water on the fun that CB'ers were having with a radio service that had been designated by the FCC as being completely utilitarian and not at all frivilous. Although this document has long been lost in the cloud of dust kicked up by the passing years, it remains, today, a totally fascinating insight into the FCC's CB policies and viewpoints at the dawn of CB-- before the agency finally came to realize that it had a panther by the paw.

FCC Bulletin #81482
7 December 1959

"The Commission has over the years established about 24 different radio services to provide for varied communications requirements of different segments of both local government and industry. Among those radio services are included the Police, Fire, Local Government, Forestry Conservation, Power, Petroleum, Railroad, Taxicab, and many others. While not specifically defining the rules of the type of messages to be transmitted, in each case the licensees are restricted to those communications considered essential to that portion of their activities which forms the basis of their eligibility in the particular service.

"In line with this policy, the Citizens Radio Service was created to provide a means of transmitting the substantive and useful messages, in connection with either business or personal activities, of private citizens who may not be eligible in any other radio service except amateur. Since the provisions of the rules governing the other radio services had not been misinterpreted, it was believed unnecessary to define in great detail the type of use which might be made of Citizens Radio stations.

"Accordingly, the service was was defined as 'A radio communications service of fixed, land, and mobile stations intended for personal or business radiocommunication, radio signalling, control of remote objects or devices by means of radio...' There was no intention to create a service parallelling the Amateur Radio Service nor was it intended that the Citizens Radio Service be used as a hobby in itself, for technical radio experiments, or for general 'contacts' of a random nature. Provision for this type of operation had already been made in the Amateur Radio Service and is available to all persons who might be interested in this type of operation.

"However, it now appears that a number of licensees of Citizens Radio stations, particularly Class D stations, have either intentionally or mistakenly interpreted Part 19 of the rules to permit unrestricted use of the stations and amateur-type use of this service on the limited number of frequencies available, and that any substantial use of Citizens Radio stations for amateur-type or 'rag-chewing' activities would create intolerable interference and defeat the Commission's purpose in establishing the service. In an effort to stem this improper use, the Commission's field offices have begun citing stations observed engaging in such activities. In addition, the Commission has issued a Notice of Proposed Rule Making in Docket 12987 which, if adopted, will more clearly prohibit amateur-type activities in this service.

"The provisions most frequently misunderstood are Sections 19.1, 19.61(s) and (c) of the rules. Section 19.1, although written in terms of a specific limitation, does indicate the intended scope of the service as 'designed to provide for private short-distance radio communications.' The Commission believes that any reasonable interpretation of the term 'short-distance' would restrict a Citizens Radio station to local groundwave coverage as distinguished from transmissions which would depend on temporary 'skip' conditions.

"Section 19.61(a) reads as follows: 'Each station in the Citizens Radio Service is authorized to communicate with other stations in the same service. Communications with stations licensed in other parts of this chapter or with any United States Government or foreign station is prohibited, except for communications relating to civil defense in accordance with the provisions of Section 19.83.'

"This provision of the rules was intended as a limitation on stations rather than a permissive key to open the service to amateur-type communications. Thus, the Commission feels, in the light of the foregoing and the whole of Part 19, that the proper interpretation of that section is that although Citizens Radio stations are intended to communicate primarily with other units of the same station, when necessary for purposeful and substantive communications in connection with the business or personal activity involved, such stations may communicate with other Citizens Radio stations. Under no cicumstances, other than civil defense activities or similar emergencies, may they communicate with stations not licensed in the Citizens Radio Service. Examples of permissible intercommunication include an exchange of messages between boats, hunting or fishing parties, or coordinated business activities.

"If a licensee restricts his transmissions to only those communications which are substantive, he will automatically comply with the requirements of Section 19.61(c) that, 'All communications shall be limited to the minimum practicable transmission time.'

"Accordingly, a licensee of a station in the Citizens Radio Service should neither call nor answer distant stations which are located outside the local groundwave coverage area of his station, nor should he relay messages from a distant station or a local station. Even within his local area he should refrain from communicating with other stations unless he has a definite and purposeful communication requirement. The practice of using a 'test' call for the purpose of inviting 'DX contacts' will be considered by the commission to be a subterfuge in lieu of the general call 'CQ' and a violation of the rules. In short, the Amateur Radio Service is available to all persons who seriously desire to engage in amateur-type activities, and the Citizens Radio Service was intended to provide the average citizen with with a means to communicate by radio when necessary

During the 1960's, all sorts of complex schemes were devised by CB'ers in an effort to stay within the FCC rules while still having fun shooting skip. One idea tried by a club calling itself the United CB'ers of America, was to issue its members "unit numbers" of the same CB license, even though there were hundreds of members spread across the nation. Unit 286 working Unit 755 might then be a legit Florida-to-Idaho contact. The FCC didn't agree and went after several such groups, including the UCBA. This QSL from UCBA Unit 909 (the group's founder) depicts the FCC in defeat. In reality, the FCC won and gave Unit 909 a considerable amount of heat.

in the conduct of his personal affairs or business activities. If you are interested in using radio in a manner which is normal in the Amateur Radio Service, you should obtain a license in that service and avoid misusing the privileges that might be granted under a Citizens Radio station license. The fact that there are over 190,000 licensed amateur radio operators in the United States indicates that the knowledge and skills necessary to pass the examination and obtain a license are not difficult to acquire."

Deaf Ears

The FCC announcement certainly couldn't have been more succinct. Nevertheless, it served no useful purpose in clearing up the problem. If anything, the service became more freeform, wild and wooly as time went on and new subterfuges for working DX evolved-- tricks such as CB'ers calling their "Unit 2" or "Unit 3," with stations hundreds of miles away replying to such calls as "Unit 2" and "Unit 3."

When it became obvious to the FCC that the 1959 memo was being ignored, the agency then embarked upon a series of Part 19 modifications and clarifications. In actuality, these were further limitations and restrictions placed on operators, each spelled out in highly specific language, complete with examples of "do's" and "don'ts. All that accomplished was to establish more rules to be violated, and inspire still more highly intricate methods of attempting to operate outside the intent of the rules while not appearing to do so.

For instance, several national clubs were formed that attempted to assign each member a unit number of the club license. All members from Maine to California became units of the club's central station. Such clubs learned, to their dismay, that their licenses (which had been issued for maybe 4 or 5 units) weren't valid to cover hundreds of stations.

All of these problems became magnified 100-fold when, in the mid-1970's, CB radio became a national fad with tens of millions of stations packed tightly on the 23 channels and spilling over the band edges onto unauthorized frequencies-- thanks to the return of good DX conditions, highly publicized CB use by truckers on the nation's Interstate highways, movies, TV, pop music, and the alleged oil shortage that made motorists want to install units

The FCC's early CB rules were in a constant state of revision.

in their vehicles to try to locate working gasoline pumps. Also, the little sets were discovered to be useful for finding police radar units, for summoning aid for distressed motorists, for asking road instructions-- plus all of the many other hobby and social benefits for meeting new friends and chatting with stations in distant lands.

As time went on, the FCC seemed to realize that it had lost the battle to cause CB radio to jell in the same shape of the mold it had so carefully planned and constructed. Reluctantly, the FCC backed away from most of its long-held restrictions. One-by-one, the regulations were shelved-- hobby use was permitted, time limits were scrapped, callsigns and individual licenses were shelved, and the number of authorized channels was increased from 23 to 40, thus legitimizing operations on former bootleg Channels 22A and 22B, also at least those out-of-band activities that were taking place between 27.26 and 27.41 MHz.

Although the FCC still vigorously enforces (within its manpower and budget limitations) violations of its technical standards (such as for power output, modulation percentage, frequency tolerance, and out-of-band operation), as well as its rules against communicating with skip stations, for the most part CB'ers are now mostly free to enjoy CB as they see fit. Of course, you can't use CB to aid in the commission of a felony, but you're invited to make the fullest possible use of the service. Millions of operators are doing just that, and it is the aim of this book to aid you as much as possible in reaping the maximum benefits from your equipment and participation in this truly unique radio service which, after several decades of struggle, has finally become the voice of the people-- truly and finally living up to its title-- the Citizens Radio Service.

To be sure, there are still those who run more power than is permitted and who operate on unauthorized frequencies. And, still, the majority of operators (from my own personal observations) can't resist shouting a distant station when the opportunity arises. The harm in such rule infractions could possibly be more in the FCC's mind rather than anywhere else, but (like 'em or not) those are the FCC's rules. This book will examine all of the aspects of CB radio and how each is being used.

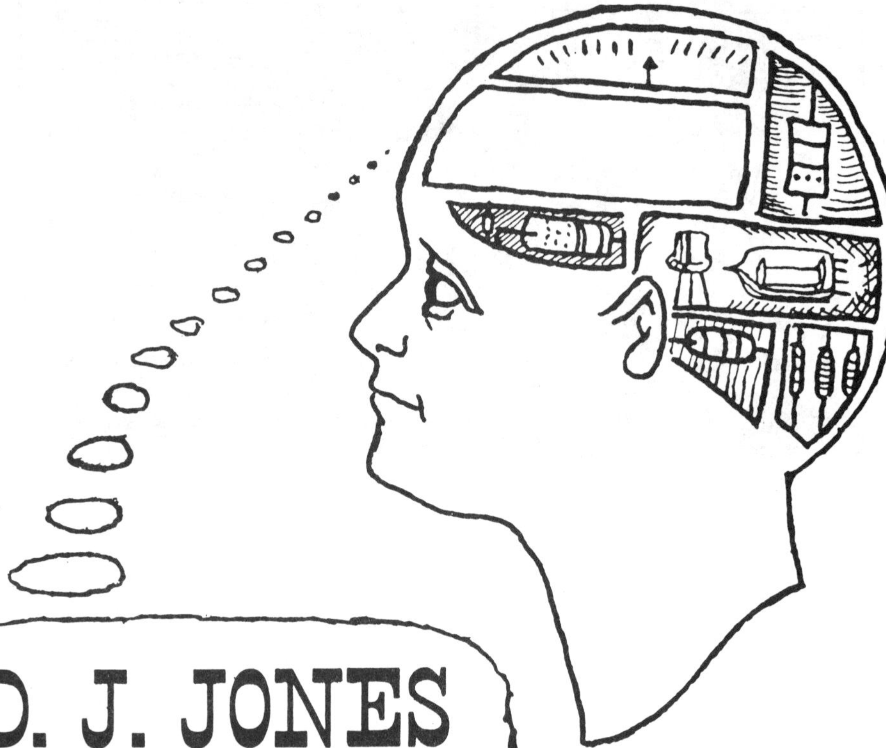

D. J. JONES

by INVADER (Chris)

This is a story I am about to tell
About D. J. Jones who is now in hell;
A man so dedicated I could write a book
Of his frequent ventures into the FCC outlook.

My story begins in a very little town
Where this man wore the DX crown,
From east coast to west, to the shores of
Singapore
He talked all night till it all was a bore.

Then one night started his plight
When D. J. Jones used his brain to its might,
With linear amps and his best turntable
He created a station using ten mile of cable.

He played his records from seven thru eleven
I swore it could be heard even in Heaven,
From the Beatle's concert to the Cosby Hour
His potent signal darted from the tower.

"Greetings CB'ers," came his dubbed voice
Thru so many channels that they had no choice
But to track him down using the loop
And beat him silly taking turns in the group.

They poured in their cars with the DF's sniffing
In front of his house, a rat they were whiffing;
A rock thru his window gave him the idea
That the lump on his head meant his listeners
were here.

They forced into his house with fiery rage
And around him made like a human cage,
With all arms jutted outward, his face
bloody red,
They let little Sally kick him square in the head.

Little is there to say of where he was born
But at least we know that his house is gone,
And behind some flowers can be read a stone
"Here Lies the Bones of D. J. Jones."

2

Identifying Your Station

For many years, the FCC assigned individual callsigns to CB stations. The first identifications issued were considered by the FCC (at least) to be little more than serial numbers, as opposed to the more formal callsigns awarded to ham, broadcast, police, ship, and other communications stations. The FCC probably felt that the flea-powered units with their short transmitting ranges needed little more than file numbers for FCC record-keeping purposes rather than traditional callsigns. These file numbers didn't even conform to international callsign regulations.

CB licenses, in the early days, could be granted by any of the FCC's 24 district offices, or you could apply directly to the main FCC office in Washington, DC. Licenses issued in Washington contained the letter "W," while those from the district offices contained the letter "A." Ahead of the letter was a 1 or 2 digit number representing the FCC district office area of the licensee. After the "W" or "A" were 4 other numbers indicating the individual station-- hence CB ID's such as 24W1607, 2A0305, 18W2101, 11A1335, etc.

Later, the FCC switched over to similar numbers, but using only the letters "A," "B," or "Q" instead of the letter "W"-- and the district offices stopped giving out callsigns. The FCC seemed to be indifferent to the problems of stations in the 4th FCC District who had been given ID's containing the letter "Q." It took a bit of courage to announce a callsign such as "4Q2367" or the like.

Then, as CB became popular and the available numbers began to be depleted, the FCC announced a plan whereby there would be additional letters added after the "Q" to permit more numbers to be issued. This caused the issuance of tongue-twisting ID's the likes of 18QA2766, 19QB5548, etc. The hope was to change the letter "Q" to a new letter each year, so that the station's ID would reveal the year in which the license was issued. Since licenses were valid for five years, these ID's were to be established on a 7 year cycle. Any modifications to the station license (early renewal, change of address, even adding units) meant cancellation of the original FCC license and ID, and issuance of a totally new license and ID-- this had always been the FCC's CB policy. A friend of mine, 2W2933, moved a few months after his license was issued and suddenly found himself changed to 2W4887!

Under the rotating system, the cycle would kick off with "Q" calls and go year-by-year until "W" calls were issued, then start over. The whole idea had to be dumped when, about 1962 or so, the FCC was notified by the International Telecommunications Union that several nations had complained that we were ripping off callsign blocs that didn't belong to our country.

No problem. The FCC then devised and switched over to orthodox callsigns consisting of 3-digit calls with the prefix "K" followed by 4-digits (such as KBG4303, KKK4758, etc.). The two letters after the "K" were to indicate the FCC District Office area in which the station was located plus the year

The early CB callsign formats were unusual, to say the very least. Mid-Atlantic operators who ended up with "4Q" often drew snickers from other CB ops. Eventually, the FCC had to withdraw these formats because of squawks from other nations who said the FCC had ripped off their callsign blocs.

in which the license was issued-- all via a highly complex series of guidelines that was a monument to bureaucratic creativity.

For a while it was confusing while the amazing conglomeration of old and new callsigns met head-on over the CB channels. For instance, in Chicago it would have been possible (in 1964) to hear stations using callsigns from no less than fifteen different numbering systems.

Eventually it was found to be impractical to attempt to issue such carefully planned and informative callsigns. The same 3-letter/4-digit callsign format was then retained but simply given out in total alphanumeric rotation without regard to date or location. The old style calls were eventually all phased out. By the late 1970's, so many CB'ers had taken out licenses that the FCC computer hit KZZ9999 and the agency decided to then switch over to a 4-letter/4-digit format (such as KBAR7694, KDCN9737, etc). Several times, during the 1970's, the FCC computer crashed and mistakenly issued the same callsign to more than one licensee-- perhaps 100,000 CB'ers were given someone else's call before the FCC spotted the error.

Much Ado About Nothing

Of course, long before things got hectic enough for the FCC license computer to suffer its several nervous breakdowns from overwork, CB'ers were no longer using the FCC's CB callsigns anyway! By then, AM-mode operators were all using code names, known as "handles," and those using single-sidenand mode were all using numbers representing affiliation with one or more networks. The use of single-sideband (SSB) identification numbers is covered in portions

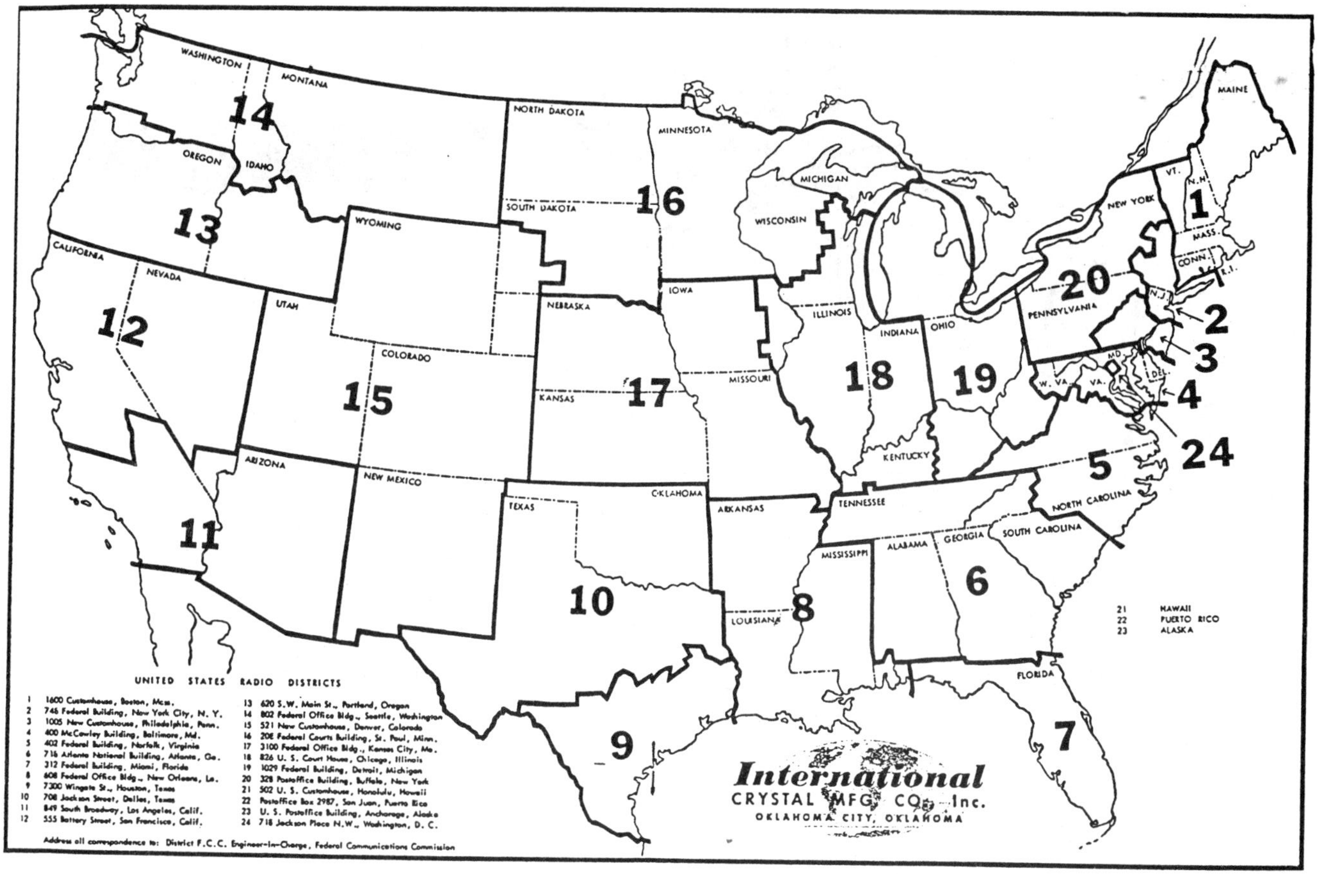

This 1960 CB-prefix wall-map used to sell for 25¢, although one in good condition today is probably worth 60 times that amount to the many CB'ers who avidly collect early-CB memorabilia. Stations in Hawaii used the prefix numbers 21, Puerto Rico was 22, Alaska was 23. It was confusing.

of this book devoted especially to SSB operation, since it is more-or-less a separate subculture active within the overall structure of the CB service, and rather far removed from the most commonly employed type of CB communications-- that using AM-mode.

CB Handles

Even though the FCC no longer issues CB callsigns, CB'ers are welcome to continue to use any such callsign they might have been issued in the past, even if the license for that callsign has expired. Doubtful that you'll hear very many stations doing so-- stations using AM-mode mostly identify with a handle.

In the ham world, a handle is the first name of the operator. In the world of CB radio, a handle is the codeword or name used to identify the station. In CB-lingo, the operator's first name is called a "first personal."

Why "handle"? The dictionary describes a handle is a part of a thing made for the primary function of facilitating the grasping of an object-- so if someone wants to get ahold of you on CB, they use your handle.

The first time I ever personally heard a CB handle used was in 1961 when one of the more popular stations heard during skip openings was the station operated by Kurm Jessop, 18B2250, of Jessop's Candies' in "Candytown, Indiana." Jessop was famous throughout the nation as the **Candyman.** It always gives me a laugh when I hear stations today claiming first rights on the handle **Candyman**! Another early handle recalled was **Bag of Pennies,** used

TOMCAT

My CB handle, Tomcat, was selected when Channel 19 came into vogue for en-route mobile use-- although all of my base station CB activities are in sideband mode where handles have never been used or tolerated by operators.

as early as 1962 by Jess Wyder, 2W6173.

Handles really didn't start becoming widely used until the late 1960's, and once they started to become fashionable, they spread in jiffy time. By 1974, the FCC's callsigns seemed to be used exclusively by newcomers for the first couple of days they were on the air when they realized that a callsign was the horrible mark of a greenhorn!

Sometimes you hear a CB'er using his actual name as his handle-- **Bill** or **Sam.** This usually raises eyebrows and seems to violate some unwritten law or code of honor on the band. After all, why should someone be just plain **Sam** when he has the option to be something exotic-- like **Sam the Man** or **Gear Jammer**? That's why I'm **Tomcat** as I roam the nation's Interstates tuned to Channel 19.

While it can't be denied that the reason that long-ago operators devised handles to thwart detection by FCC monitors, in later years most folks seemed to use handles as an extension of their own personality or some image of themselves that they wish to project to the world. This has been the cause of no small amount of problems.

And still, if you think about it, such personalized names are really nothing new-- names are given to boats, cottages, and lots of other things. People have nicknames that present the proper image they have of themselves. Perhaps your name is Richard, but folks call you Rick, Rich, Dick, Dickie, Ricky, Richie, or Slim. If so, you've already used an alternate name (or handle) for years before you discovered CB and started racking your brain about something special (and different) to call yourself!

It may be possible to come up with a CB handle that's unique, but you may have to use your imagination and find one that is so odd that nobody else would think of it-- or so dumb that nobody else would want it.

When a CB handle is selected (sometimes after much thought) it becomes something so personal that its owner likes to consider it as his or her private property. A person who pondered long and hard to come up with the handle **Lone Stranger** may well consider it to be a major affront to flip on the CB rig and immediately come face-to-face with another **Lone Stranger.** I mean, how do you think Hulk Hogan would feel if he climbed into a wrestling ring and found that his opponent was yet another guy calling himself Hulk Hogan? It just isn't done! Right? Suppose there were two radio broadcasters in your area calling themselves, "The Wonderful Voice of West Hurleyville"? Just not done!

So, after carefully considering and rejecting dozens of possible handles, you and your family/friends agree on one that captures the real you-- you may therefore be both startled and offended to find that the first time you use it on the air that someone else in your area claims to have used it first is is enraged to find that you have apparently ripped off their exclusive identification-- have somehow attempted to steal their identity and possess their reputation and personality. You may find that the theft is regarded as seriously as if you had stolen the CB from beheath their car's dashboard. CB'ers have actually come to violence over the alleged theft of their handle by some other operator. A person who has spent years becoming famous as **Deacon** suddenly finds that there's someone else on the radio masquerading under their very own handle. It's spooky and almost supernatural. The victim of the theft lurks around the edges of the band-- a hollow shell-- as someone else (probably unknowingly) appears to bask in their reputation. Sounds like something from a story by Gogol or Kafka!

Of course, before it comes to this, there is often a cordial warning given-- "Say fella, how long ya been usin' dat handle? I been goin' by da handle **Blue Streak** for 36 years now!" This is a definite warning shot fired across your bow and should be instantly regarded as such.

Your choice is to say, "Golly, sir, I'm awfully sorry-- I'll pick another handle right away!" Or, you can take the "So what are you gonna do about it? That makes two of us!" approach. One clever retort is, "Oh, you're just the guy I've been looking for-- I've been **Blue Streak** for 40 years and I've been hearing about some guy who ripped it off!"

Play it by ear. Remember that there are only a finite number of words and 2 or 3 word combinations in the English language that can be used as CB handles-- and there are already many times the number of ops than those potential combinations. So, unless you decide to come up with a handle that nobody else will want or could possibly have thought of-- like **Neon Toenails** or **Grebny Fiddleknocker,** you can be pretty certain that the handle you're using is also used by many others who came before you-- and who will come after you-- especially if you've picked a popular handle like **Blue Knight, White Knight, Bandit, Foxy Lady,** or **Gear Jammer.** I've probably met up with a sufficient number of other guys using the handle **Tomcat** to fill a Boeing 747.

Sometimes operators change handles without a second thought, flitting around to different ones on a regular basis because they never feel any real affinity with those selected. Other operators knew the one correct handle for their use even before they purchased a rig. To still others, the selection of the proper handle is an exercise in metaphysics, which includes meditation, seeking out the right "vibes," and asking friends, Ouija Boards, and astrologers!! Personally, I've always felt the best method is by monitoring the band until you hear a good one, and then swiping it. That's how I picked **Tomcat** as my own handle.

If you'd like some help in picking a handle, permit me to present some suggestions chosen from the many different categories of handles. For instance, you might like a job-related handle. Like a pharmacist using the handle **Pill Roller,** a mechanic called **Valve Job,** an attorney whose handle is **Legal Eagle,** a nurse who is **Molly Pitcher** or **Florence Nightingale.** You get the idea. Try guessing the occupations of CB'ers using handles such as **Teacher's Pet, Printer's Devil, Shutterbug, Sawbones, Farmer Alfalfa,** and **Scribbler.**

You hear many hobby tie-in handles-- like the fishing fanatic who goes by the handle **Rod 'n Reel,** the chess player who is **Checkmate,** the camper who is known as **Tent Peg,** and a coin collecter of my acquaintance who prefers the handle **Reeded Edge.**

How about handles with a geographic tie-in? Like **Texas Kid, Connecticut Yankee, Gas House Charlie, Misplaced Mexican, Philadelphia Phil, Bayou Billy, Montreal Millie,** or others of that ilk.

Sports-related handles are often heard, too, such as: **Louisville Slugger, Backstroke, Tennis Elbow, Boots 'n Saddle, Skydiver, Skipole, Jogger,** etc.

Lots of handles connected with the color or type of vehicle owned by the CB'er-- **Heavy Chevy, Blue Blazer, Bronco Billy, Dan in the Van, Van Man, Blue Streak, Green Hornet, Vette Set, Rollin' Thunder,** etc.

CB handles all reflecting an "on the go" motif. Big Mama is the nickname of the large truck driven by a Dutch operator. The Copper Caboose reflects SSB-3036's interest in railroading. Little Bandit was probably selected by SSB-2932B in homage to the film Smokey and the Bandit.

Don't overlook the appeal of handles connected to movies, TV, music, and entertainment industry personalities, such as: **Captain Kirk, Betty Boop, Rolling Stone, Disco Gal, Snoopy, Spock, The Boss, Sledge Hammer, Max Headroom, Iron Maiden,** are a few I've come across.

A popular category of handles projects a somewhat provocative image: **Cuddles, Foxy Lady, Hot Lips, Passion Flower, Macho Man, Lover Boy, Dapper Dan, Beach Baby,** and many other similar gems.

Let's not overlook the ghoulish spookshow category, like: **Dracula, Frankenstein, Fang, Phantom, Exorcist, The Claw, Dr. Blood, Ghostbuster,** etc.

There are countless astrological sun signs-- **Pisces, Aquarius, Gemini, Sagittarius, Scorpio,** and the like.

CB'ing police officers sometimes select handles like **Kojak, Diesel Chaser,**

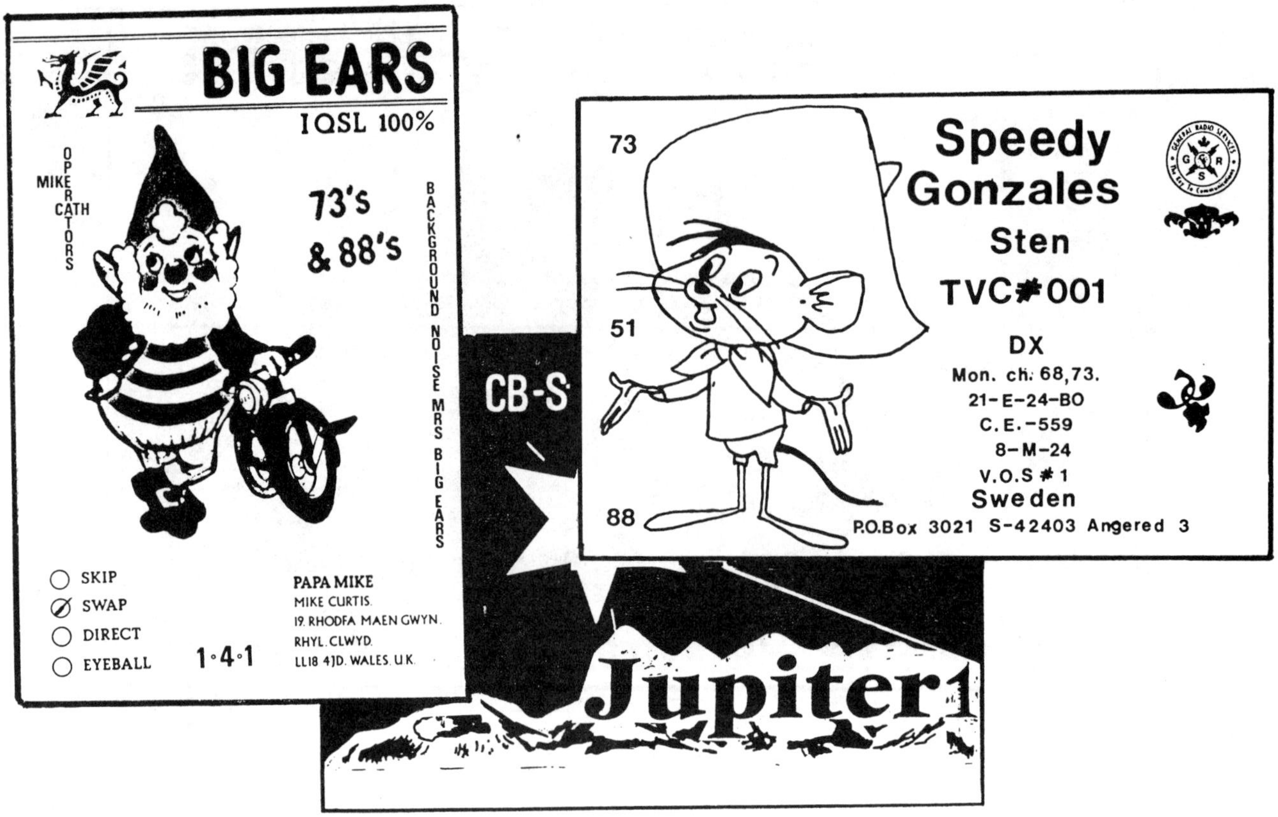

Greenstamper, Big Bear, Smokey Joe, Big Badge Man, Picturetaker, etc.

Well, you get the picture, and you can use these ideas to prod your own imagination in order to come up with something suitable. There are actually dozens of other categories, such as those with ethnic connotations (**Italian Stallion, Black Velvet,** etc.), ones preferred by kids, those that refer to some outstanding or unusual physical characteristic or attribute, pavement princess handles, handles that are based upon the name of the operator, drug related handles, animal and pet related ones, nonsense words, agricultural references, political ones, puns, and so on-- including crossovers and combinations.

When selecting a handle for ones self, or encountering someone else's handle, pay attention to things such as if it's self-humiliating or self-degrading, like like **Jelly Belly, Bullet Head, Big Dummy, Sad Sack, Wretched Roger,** or **Big Butt** , or if the handle is either aggressive like **Superman, Bone Crusher, Big Bruiser, King Kong, Mighty Mouth, Number One,** etc., or super passive-- **Wallflower, The Listener, Me Next, Mr. Quiet, Little Flower, Cream Puff** are ones I've heard recently.

My guess is that you can probably tell more about a person by his or her CB handle than from a dozen Freudian shrinks aided by a convention of astrologers and crystal ball gazers. Think about it, a CB handle is possibly no more or less than a capsulized summary of what a person is. Better make sure that the one you select doesn't project the wrong image, or let any skeletons out of the ol' closet.

So there you have some thoughts on the subject of CB handles. Before you key up that mike button (again or for the very first time), spend a few extra moments deciding if you really want to be thought of as **Horseface, Mr. Miserable, Silly Seatcovers, Snaggle Tooth, Tightwad,** or something of an equally revealing nature.

For those who operate on the frequencies normally used by SSB operators, none of this should be a problem. The operators there don't tolerate the use of handles. Showing up and announcing a CB handle on a SSB frequency will make you as welcome as the Wicked Witch was in the Land of Oz.

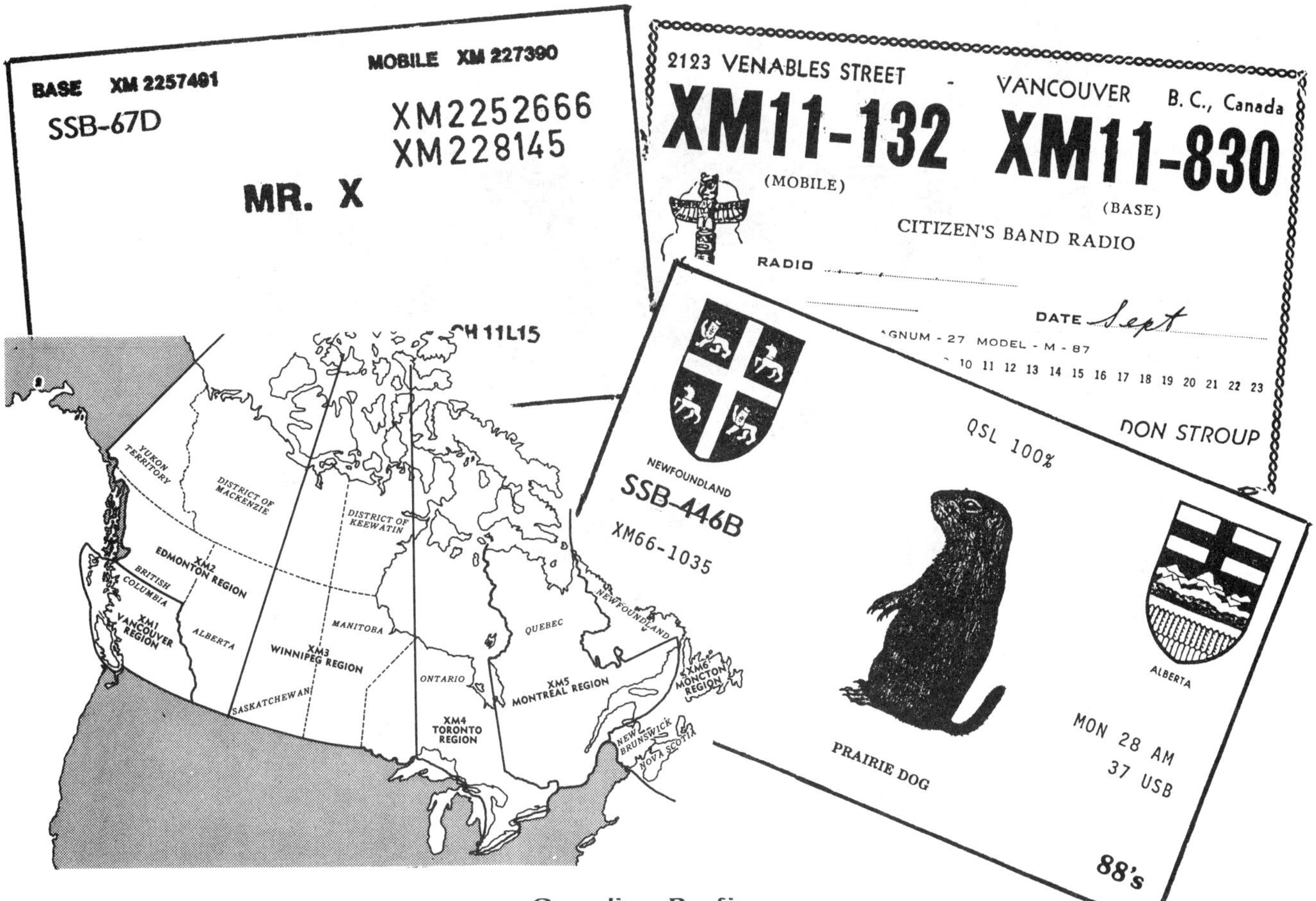

Canadian Prefixes

Callsign prefixes of Canadian stations reflect the general area where the General Radio Service license was issued, either by a government field or regional office. A callsign is issued with the prefix letters "XM" followed by two numbers between 11 and 66 that indicate the office where the license records are maintained. This is followed by additional numbers issued to the individual licensee. Therefore, a call such as XM42-47386 would be from the area of the Toronto, Ontario, field office.

The prefixes used are:

XM11 Vancouver, BC (Regional)
XM12 Vancouver, BC (Field)
XM13 Victoria, BC (Field)
XM14 Prince Rupert, BC (Field)
XM15 Kelowna, BC (Field)
XM21 Edmonton, Alta. (Regional)
XM22 Edmonton, Alta. (Field)
XM23 Calgary, Alta. (Field)
XM24 Grande Prairie, Alta. (Field)
XM31 Winnipeg, Man. (Regional)
XM32 Winnipeg, Man. (Field)
XM33 Regina, Sask. (Field)
XM34 Saskatoon, Sask. (Field)
XM35 Brandon, Man. (Field)
XM36 Port Arthur, Ont. (Field)
XM41 Toronto, Ont. (Regional)
XM42 Toronto, Ont. (Field)
XM43 Hamilton, Ont. (Field)
XM44 London, Ont. (Field)
XM45 Kitchener, Ont. (Field)
XM46 S.S. Marie, Ont. (Field)
XM47 No. Bay, Ont. (Field)
XM48 Kingston, Ont. (Field)
XM49 Ottawa, Ont. (Field)
XM51 Montreal, PQ (Regional)
XM52 Montreal, PQ (Field)
XM53 Sherbrooke, PQ (Field)
XM54 Pt. Alfred, PQ (Field)
XM55 Quebec, PQ (Field)
XM56 Three Rivers, PQ (Field)
XM61 Moncton, NB (Regional)
XM62 Moncton, NB (Field)
XM63 Halifax, NS (Field)
XM64 Sydney, NS (Field)
XM65 St. John, NB (Field)
XM66 St. Johns, Nfld. (Field)

Canadian GRS operators normally use CB handles for AM-mode and sideband ID numbers on the SSB frequencies.

The Untied States of America

★

Whereas

THE OFFICE OF THE COMMISSIONER OF PATENTED IDEAS HAS DULY RECEIVED AND PRINCIPALLY EXAMINED THE REQUEST OF

____

WHO HAS APPEARED BEFORE THIS OFFICE AND HAS PRESENTED A PETITION TO BE DECLARED THE SOLE ORIGINATOR AND INVENTOR OF

CB RADIO

NOW, THEREFORE, AFTER EXAMINING THE STATEMENTS AND CLAIMS CONTAINED IN SAID PETITION, THE OFFICE OF THE COMMISSIONER OF PATENTED IDEAS, ACTING IN ACCORDANCE WITH THE LEGISLATIVE POWERS INVESTED THEREIN, HEREBY GRANTS SAID PETITIONER THIS PATENTED IDEA SIGN ON THIS ORIGINAL INVENTION, AND HEREWITH AUTHORIZES EXCLUSIVE RIGHTS TO THE USE OF THIS INVENTION FOR A PERIOD OF FIVE YEARS FROM THE DATE AFFIXED HEREON.

Dated in Washington, D.C., on the eleventh day of September in the year one thousand nine hundred and fifty-eight.

Tom Kneitel

PATENT DOWN THE HATCHES

3

Talking on CB

There are several categories of talking on CB. AM-mode operations are predominant on the CB channels-- the AM mode is what made CB famous and what generated the enormous popularity of CB. Sideband operators are a distinct subculture on CB, and the CB rules of the road they use are vastly different than those used by the majority of operators on CB, so SSB operations are covered elsewhere in this book.

Within the scope of AM-mode operations there are hobbyists, emergency service teams, business users, truckers and other mobile units on the nation's highways, youngsters, and a myriad of other users. With such a diverse spectrum of operators on the channels, the trick is to work things out so that everybody can pursue their own primary interests with a minimum of friction and a maximum of effeciency and enjoyment.

For starters, recognize that (although it is often abused) Channel 9 has been designated by the FCC only for emergency communications or for traveler assistance. Any other use of Channel 9 is a violation of the FCC regulations-- also it puts such operators in line for causing interference to those who need to use the channel to summon help during a highway emergency. Lives could be lost-- and there are plenty of other channels available for your enjoyment. Remember that just because you can't hear any activity on the channel, it doesn't mean that your transmissions aren't jamming operations beyond the range of your receiver, or that your signals aren't being carried by skip propagation to some distant area where emergency activities are being coordinated on Channel 9.

Channel 19 is (unofficially) the frequency that is most frequently used by enroute truckers and motorists for general information exchange, smokey reports, traffic situation condition reports, and similar. As such, it's best used as a place exclusively for mobile units. With the exception of base stations established solely for the purposes of offering brief bits of information to enroute mobile units, it's certainly not the channel for general chit chat between base stations or routine base-to-mobile conversations. In some areas of the nation there are additional or secondary mobile enroute channels used on roadways that lie so close to roads where Channel 19 is normally used-- the same courtesies would apply to those byways.

Although youngsters have now decided to sprinkle their communications over many channels, Channel 14 has traditionally been avoided by many operators because it's generally regarded as the place where the little kids congregate.

Other channels that are often reported as being popular in many areas for special interest groups include Channel 8 for agricultural interests, Channel 13 for maritime purposes. SSB stations, on a nationwide basis, presently operate on Channels 34 through 40, with some areas still using the old SSB hangout of Channel 16. Since AM and SSB operations don't mix well on the same channels, AM operators should avoid operating on channels used by

SSB operators in their local area-- just as SSB operators extend AM operators the courtesy of staying fully clear of channels normally used by AM operators. It makes for better communications for everyone.

AM Station Operating Techniques

Operating techniques on AM frequencies seem to vary somewhat from place to place. It's mostly a matter of spending a day or two monitoring around on different channels before you attempt to transmit, and then governing your operations with courtesy and common sense.

Some people don't realize that they may be making instant enemies when they key up their mikes-- they want to contact a friend and they push the button and go straight into, "Breaker, breaker, breakity broke-- How 'bout ya **Slingshot**? Ya got the **Panther** calling collect. Gotcha ears on? C'mon, c'mon ya cottonpicker!"

It's an all too typical start for a CB conversation. **Panther** may have been either lucky or unlucky when his friend came back with, "Go ahead, **Panther,** you old rascal. We be accepting your charges, come on."

As their conversation progressed, others on the channel were fuming. There were stations using the channel when **Panther** decided to shout his pal. They probably only required a few seconds more to complete their chat when **Panther** barged onto the channel and stepped on the other stations.

"Come on back, **Casanova,**" urged **White Rock,** one of the two whose contact had been messed up, "They walked all over you."

It took only a few minutes for **Casanova** and **White Rock** to end their conversation, but when they finished, **White Rock** stayed on the channel to transmit a dead carrier while **Panther** and **Slingshot** were talking. His only goal was to deliberately jam their ability to talk.

"Let's take it to two-two," shouted **Panther** under the dead carrier, "We got some clown on that don't want to let anybody else talk." And neither **Panther** nor **Slingshot** understood why.

It's not an uncommon situation. Most CB'ers don't go out of their way to interrupt someone else's conversation. They're good folks who want to talk to a friend and don't understand that what the're doing is impolite.

To many CB'ers, the words "break" and "breaker" are the accepted ways of getting on a channel-- although the rights and wrongs of their use are all too vague.

"Break" means, "I know that you're using the channel, but I'd like to jump in and join your conversation or else call a friend here." It's used in the same way the phrase "excuse me" is used when two people are talking. A "Quickie Break" (or "Shortie Break") seeks to interrupt for only a few seconds to ask a question or contact your friend with instructions for switching to an unoccupied channel. The interrupted station will then usually ask the breaker to go ahead, or possibly request the breaker to stand by for a minute. Courtesy demands that the breaker follow those instructions, just as it also considered bad form for the other stations to drag out their contact when they know others are anxiously awaiting the use of the channel.

If a channel has been monitored for at least 15 seconds and is obviously not in use, then there is absolutely no reason to say anything about or ask for a "break." You simply go ahead and make your shout without any further formalities or courtesies. People who continually repeat the words "Break, Break" or "Breaker two-six" (or whatever channel) on an otherwise quiet frequency, refusing to go ahead with their shout until some unknown voice comes out of the ether to grant them permission to go ahead and use their CB rig simply don't know how to talk on CB!

If the channel is stony silent and you feel like talking to someone-- anyone-- who might be listening there, all you need do is key up the mike and announce something like, "Is anybody on the channel today? Here's **Big Mac**" (or whatever

THE ETHER HOG

"Oh, I am the hog of the air,
Wherever you tune, I am there;
I am the prize squeaker,
I fill your loud speaker—
The ether is free. I don't care.

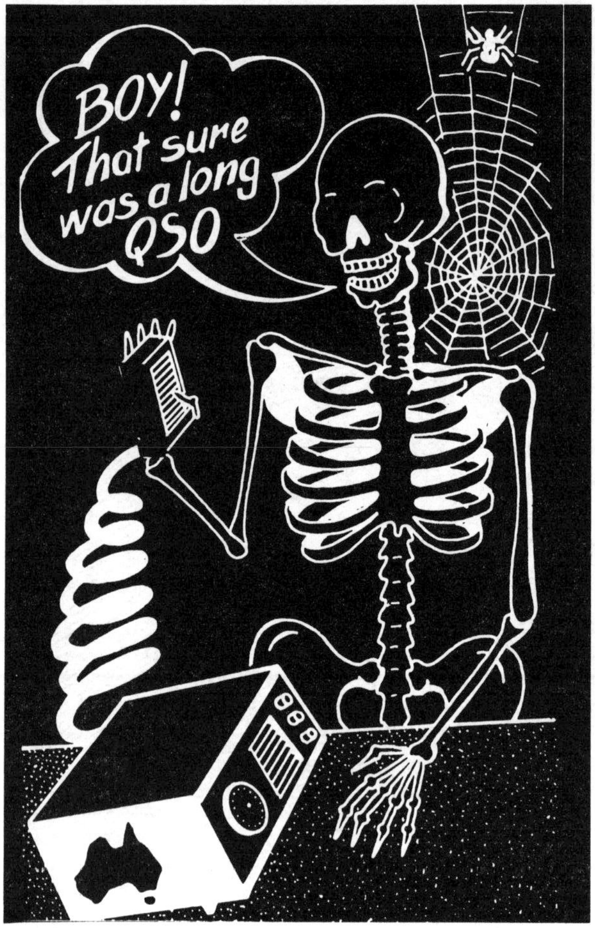

your handle is). But, for heavens sake, don't go into that silly "Breaker" routine.

Regardless of how much you try, chances are you're bound to accidentally step on somebody's contact on a rare occasion. Unfortunately, it will be your luck that it is a guy whose handle is **Obnoxious Ollie.** If and when that happens, keep your cool-- say you're sorry. Arguing over the radio or seeking revenge makes life difficult for everybody on the channel. Moreover it causes you to look rather foolish to everybody listening to the fight, especially if you either get freaked-out or have to eat crow because the other station threatens to locate your station and punch your lights out.

If CB is your hobby, keep in mind that you're supposed to be enjoying yourself and sharing the enjoyment with other hobbyists. If everything that happens on the band, and/or everything you say or do, causes friction and annoyance, fighting, high blood pressure, ulcers, and stress-- then you might wish to consider the possibility that your approach or attitude needs a substantial change. Or, maybe you might consider getting into photography or arm wrestling.

CB = Crashing Bore? No, Not You!

You may find that other CB'ers are waiting in line to hook up with you on the air. On the other hand, you might find that your contacts are those dreadful little "short-shorts" or "quickie shouts" in which the other station backs out of the contact very pronto with excuses about being called to the telephone, to answer the doorbell, to the supper table, or the bathroom. Let's hope not, because it could well be that while you are wonderfully courteous to other operators, you are also a very dull person to talk to. Could happen, be you an AM operator or an SSB'er.

Worst part about it is that persons so afflicted are invariably unaware of the condition, not having even the slightest inkling that they are wearisome, tiresome, uninteresting, monotonous, boring, and somniferous. Some of these poor souls may even think that they're amusing and as interesting as hell every time they fire up the ol' rig. And just who is there to let them know the awful truth-- it's hardly something most folks would just come out and say over the air even if the conversation lasted long enough for them to do so. Although, I do recall recently hearing one operator bluntly tell someone, "Look, fella, this shout is just going in circles-- let me back out before I doze off!"

The pity of it all is that CB is such a wonderful way of chatting with people and perfecting the lost Art of Conversation. CB has the potentials for opening up lines of mutual communication and friendship between peoples of all ages, nationalities, ethnic groups, religions, from different neighborhoods, and all other real or imagined barriers that tend to separate us. And so many CB'ers are frittering away this potential without realizing it, or knowing how easy it is to change things for the better.

Grab The Dull By The Horns

Not everybody can be David Letterman or Eddie Murphy, it's true, but chances are that a considerable amount of dull chatter could be eliminated if more operators would take a little time out to analyze their on on-the-air conversational habits and then decide whether or not they are among the blameworthy-- or, as so devistatingly put by a CB'er who calls herself **Blunder Woman,** "persons in need of a hearing by a Jaw Crimes Tribunal."

OK, so let's see what seem to be some of the major causes of tedium and then suggest ways of replacing them with alternative approaches to 27 MHz conversation that may be more rewarding for everybody concerned.

For instance, what about those operators whose chatter consists 98% of simply feeding back to the other station all information from their last transmission? You've heard it, "Yeah, 4-10 on your transmission. OK on your new beam and how well it stood up in the storm yesterday, and fine too on needing to get some new wallpaper printed, and I agree with you on that station over in Dudleytown that ties up Channel 3 too long. So I guess I turn it back over to you again for your comments-- pick it up!"

What exactly is the other station supposed to say when all there is to work with is a feeble replay of his own words. This is a contact that was doomed from the start-- a shame, too, because there was so much raw material placed on the table from which to develop into a reasonable conversation. Instead, it just laid there and died a horrible death!

Perhaps of equally boring impact are those operators who don't really have anything to say so, instead, they "vamp" (fill time) by means of long strings of passe and meaningless repeats of corny CB-type words and phrases, the total meaning of which is rather vague and ill-defined, and which would be universally suitable (or unsuitable) as a response to just about anything that might have been said. Example: "Mercy, mercy, you cottonpicker, that was a bodacious trip and, fer sure, your radidio is definitely clear as a bell, so, you old rascal, you're really hittin' me with nine pounds, blowin' plenty of smoke, cleanin' out my ears, bending the pin on the meter, rattlin' the windows like thunder in the east. So I'll turn it back to you for any comments you might care to add, and if I can't copy your next transmission I hope I catch you again sometime. And may you have a good day today and a better day tomorrow." What could possibly said in response to that monument to monotony? Exactly nothing!

Then, of couse, there are bores who actually do attempt to whip up a conversation. Problem is that their selection of topics is bad news. Think about whether you have ever put some other operator through the agonies of enduring your thoughts about the following dreary or overworked topics: mothers in law, the wife's bad coffee/cooking, your poor health or that of your spouse, inflation, taxes, the bad weather, miseries connected with your job, miseries connected with your unemployment, and those three ultimate sure-death conversational CB discussion topics-- your political, religious, or racial viewpoints and philosophies.

Many female operators complain that they get rather tired of finding that so many guys who shout them automatically assume that the only reason they have a CB rig is to meet potential studs. Unfortunately, there are ladies who have gotten into CB for the purpose of meeting guys. This is especially true of the teenie-bopper crowd. I can't think of too many things more irritating than having to listen to the cloying chatter of an all-too-cute young lady who is attempting to capitalize on this singular weakness of so many macho operators. You've heard it too, "How come ya didn't shout me for the last two days, Eddie-- doncha love me any more? I always shout you when I hear ya on the channel." Arggggh!

Hey, I don't want to be accused of attempting to kill off some of CB's most time honored and cherished traditions, those quaint things that made CB the butt of so many jokes on TV, created no less than fifty different CB magazines, sold millions of dollars worth of movie tickets, records, board games, bumper stickers, pins, buttons, and other novelties. Still, I've come to realize that there are many operators (especially newcomers) who not only don't realize that much of the "I was just a-tellin' maw" type of talk stopped being funny, clever, interesting, and even tolerable about 1980 or so. Even that famous old CB term, goodbuddy, has not only become worn out, it is presently read by many operators as a euphemism for "gay." Well meaning newcomers who don't realize this could end up with some problems should they use that form of address to an operator without a sense of humor.

As you can see, there are many ways of doing it wrong. But, what then are the things upon which successful 27 MHz conversations are built? There are ways of avoiding the pitfalls of being dull on the air. Of course, remember that it takes two or more operators to make a conversation, so you're going to have to listen around for others who, like yourself, are striving for something more than dull fare. While there does seem to be more of an effort towards meaningful conversation on the SSB frequencies than on the AM channels, don't count AM out. There are many fascinating people on the AM channels who'll really fascinate you if you give them half a chance!

Modulation Education

Listen to the channels and pay special attention to those operators you find particularly interesting, and also particularly dull or irritating. List their best and worst points on a piece of paper, and keep it by your rig for ready reference as you communicate. Eventually you'll have the list memorized.

An amazing education can be had by the simple trick of making an audio tape of an hour or two of your own CB operations. You'll get a great idea of how you sound and come across to others. Wait a few days after you make the tape before you listen to the tape, then pretend you were monitoring two other stations in contact. Is that you? Good grief!

Here are some pointers:

Avoid speaking in a dull monotone-- a bored and endless droning. Animate your voice with expression and inflection. Don't be afraid to laugh, and stay clear of saying "hi-hi" if you wish to express humor. A genuine laugh is both appropriate and welcome.

Start off your conversation in an upbeat mood, and keep it on that level. Don't bellyache over your personal problems because they are probably of

WORLD OF A CB'ER

by KRD2322

With one hand on the mike
The other on the dial—
10-4s and 10-20s flying real wild—
With condensers and transistors
Capacitors and resistors—
Crystals and channels
And come back with your handle—
Air waves, ground waves, antennas and skips
Sure, we'll check you out when you start on your trip.
Squelches, and variables, S-meters and DBs
If your signal is weak, shaky go your knees.
With 10-8s, 10-5s, 10-6s, and 10-7s
To most folks it's Greek—To a CB'er it's heaven.
So use your rig with particular care—
Or Fox Charley Charley may soon be here.
He's armed with pink tickets and show cause why—
And you better be ready with a quick reply.
So use your rig with particular care—
Make real good use of your CB Band.
Good luck and best wishes to you, you, and you,
It's 10-7 time for KRD2322!

little interest to anybody else. Don't complain about or put down other operators, local clubs or stores, various products.

Listen carefully to other operators and to what they have to tell you. Make notes on a handy pad of paper to jog you memory about the other operator's name, handle or sideband number, location-- and refer to this information as you speak. Use the other operator's name as you speak, people like that. Isn't it annoying when you tell the other operator your name's Hank, and he comes back and calls you Harry, Herb, Harvey, or Herman?

If the other operator tells you he just got a new rig or antenna, or a new set of wheels-- he's telling you to share his good news with you. Be enthusiastic in your response, don't let it slide by. Even if you think that the car or radio is a hunk of junk, tell him that the thing is fantastic. If you can't bring yourself to say that, then at least say that you don't know much about it and perhaps he could give you a rundown of its features. Don't drop the ball on bits and pieces of information being fed to you.

Some good starter topics to build a conversation include your own interest in CB'ing and how long you been involved-- what you like best about CB and some particularly useful things you've found to do with CB radio. Talk about your other hobbies, your recent or planned vacation, some humorous aspect of the daily news (if you can find anything funny in what's going on). Don't be afraid to ask questions if you feel the other operator needs a little revving up to become fully conversational.

Don't ruin an interesting conversation by dragging it on and on, far past the point where it's apparent that it has run its course. Maybe next time you and the other station will have something else to discuss. Besides, others may be waiting to use the frequency. In fact, if your conversation looks like it's starting to become a bit lengthy, it's considered good form to pause from time to time and ask if there are any breakers who want to use the channel-- you can always pick up your chat when they've finished their shout.

If you happen to tune in on two other stations in conversation, don't kibitz anonymously from the sidelines with snide comments and wisecracks to horn on their shout. If you want to join in, ask for a break and you'll probably be welcomed. Roundtables consisting of three or more stations can be lots of fun and are excellent for breaking the ice for inexperienced operators who are slightly bashful or suffer from a touch of mike fright.

You'll probably find that in your local area there is one popular channel that most stations monitor. While that channel is fine to monitor, chances are that if you've got a long-winded conversation on tap, other area operators will most likely appreciate your courtesy in taking it to another channel and leaving the local monitor channel open for calls and quickie contacts.

Above all, don't worry about having a touch of mike fright at first. Just recall that everybody remembers having that sinking feeling upon first pushing the mike button. Most CB'ers are people just like you, folks who want to meet you and welcome you into their fraternity. Hopefully, we've given you some ideas on how to do it in style.

One last thing, CB'ers generally avoid announcing their last names, street addresses, or telephone numbers over the air. Probably dozens of reasons why, but we suggest not asking any other operators for such information nor making yourself look like an oddball by providing such information about yourself over the air. Usually, however, post office box number addresses are considered acceptable for over-the-air exchange.

CB Lingo

What eventually became famous as CB lingo and jargon came into being around late 1974, and within only two years had evolved into a huge array of words and phrases that were recognized by the general public as being part of the mystique of CB radio. Seemingly, there were CB words and phrases for every possible communications or vehicular situation, special CB nicknames for many cities, pieces of CB equipment, types of vehicles, and whatever.

CB jargon inspired films and TV sitcoms, plus a seemingly endless parade of songs from the tune factories of Nashville. If the truth be known, a great deal of the so-called CB jargon was first invented by Hollywood and Nashville and made it briefly onto the CB channels only after CB'ers and everybody else had the words drummed into their heads by the national media which was riding high on the CB fad in the mid-1970's. Although there were supposedly more than 2,000 CB words and phrases in existence at the peak of the mania, most were never used by the majority of operators. Indeed, much of this so-called CB lingo probably would not have been understood by CB operators since it consisted primarily of media-created artificial words and phrases that amused the public.

By now, the bulk of this has been filtered out by time. Still, there are words and phrases that encountered, especially on Channel 19. Some words are too localized to be encountered on a national basis, but others do seem to be floating around from coast-to-coast. Sidebanders seldom use these words and phrases, but those operating in AM mode may well encounter them, especially from those who have been active on CB for several years.

Here are some of the words and phrases you might encounter:

Ace: A hairy-legged old skipper who fancies himself a hotshot operator.
A.D.: An operator's mailing address. (AM/SSB use.)
Advertising: A marked police car with lights/siren going full blast.
Afterburner: A linear amplifier.
AM'er: A CB'er who operates mostly/only in AM mode.

What'd you say the 20 was on that Evel Knievel, pal?

Ancient Mary: AM mode.
Ancient modulation: (Ditto.)
Any eastbound (westbound, etc.) mail? Requesting traffic/smokey information of use to vehicles heading in that direction.

Back Door: Last vehicle in a convoy checking for approaching smokies.
Back down: Reduce speed ahead.
Back in a short-short: Leaving the air briefly. (SSB = QRX)
Back it on down: Reduce vehicle speed.
Back off the hammer: Reduce vehicle speed.
Back out: To sign off from a contact.
Backstroke: Return leg of a round trip.
Barefoot: An unmodified CB rig without a linear amplifier.
Basement: Unauthorized freqs. below Channel 1/26.965 MHz.
Bear: Any police officer on highway duty.
Bear cave: Police station. (Also "bear den.")
Bear trap: A police radar speed trap.
Beat the bushes: First vehicle in a convoy watching for smokies.

Big 10-4: Signifying emphatic agreement.
Bleed: Hearing signals from the adjacent channel. (Also "bleedover" & "splash.")
Bodacious: Corny/cute way of saying strong, great, wonderful.
Boots: A linear amplifier.
Boulevard: A limited access or Interstate highway.
Break: Request for permission to interrupt/join someone else's contact.
Breaker-Breaker One Nine: What big dummies and rank greenhorns enjoy saying when the want everybody on Channel 19 to know what channel they're on, just in case they forgot. In exchange for providing this information, they beg your permission to use the channel. Occasional use on other AM channels.
Breakity-Break-Break: A supposedly humorous way of breaking.
Bring it on back: I'm waiting for your reply.
Brush your teeth and comb your hair: Approaching a radar trap.
Bubblegum machine: The red light atop a police car.

Bushel: 1,000 lbs.
By: I'm monitoring the channel for any calls.
Bye: Terminally cute way of signing off. (Also "bye bye.")

Camera: Police radar.
Capabilities: Adding extra frequencies or a slider modification to a rig.
Catch you on the backstroke (or flip-flip): Talk to you on the return trip.
CB'er: A person who operates a CB station.
Channel hog: A CB'er who talks so much that others can't use the channel.
Channel monitor: Someone who takes it upon himself to control a busy channel by letting waiting "breakers" know when it's their turn to talk. Most take delight in this duty and eventually become mini-dictators-- turning "control of the channel" over to their cronies when they short break or sign off for the night. More often than not, this job seems to be the emotional release for the severely henpecked or the poor soul who is at the bottom of the totem pole at his job. All persons hoping to use the channel are expected to patiently "wait in line" as the benevolent monarch advises, "You're Breaker #4," or some such nonsense. Thankfully, this is less common now than it was in the 1970's.
Channelmaster: A CB'er who feels he personally owns a particular channel.
Charlie: The FCC.
Chase car: The police car sent out to catch speeders detected by a radar trap.
Chicken coop: A truck weighing station.
Chucker: One who causes deliberate interference with a dead carrier,
Clean and green: No police cars in sight.
Clear: Contact completed.
C'mon, c'mon: Invitation for another station to transmit. For reasons still unknown it is usually pronounced something like "kay-moan, kay-moan."

Coffee break: An informal CB gathering, often at a diner or parking log.
Convoy: Three or more vehicles traveling in single file.
Copy: To copy is to listen to another station. (AM/SSB use.)
Cotton in your ears: Your receiver can't hear anything.
Cottonpicker: For many years, the foulest possible epithet one CB'er can call another. Now, mostly used in joking between friends. Probably best to not try using it on anybody you don't know very well. The SSB equivalent is "lid."
County mountie: A deputy sheriff or county police officer.
Cowboy Cadillac: A Chevy El Dorado, or a fancy Blazer/Bronco or other 4X4.
CQ (or CQ-DX): Calling any distant station-- please reply! (Mostly SSB use.)

Dead carrier: An signal without anybody speaking.
Do it to it: Driving at full speed.
Don't feed the bears: Slow down, radar ahead.
Double nickels: 55 m.p.h. (Also "double nickels on the board.")
Down and gone: Contact ended.
Downstairs frequencies: Unauthorized freqs. below Channel 1/26.965 MHz.
Dragon wagon: Tow truck.
Driver: Replaced "goodbuddy" as the term to call another mobile operator.
Drop the hammer: Accelerate to top speed, road is clear.
DX: A distant station heard via skip.
DX'ing: Unauthorized 2-way contact with a very distant station by skip.

Ears: The receiver portion of a CB transceiver.
Eatum-up: A roadside diner, restaurant, or truck stop.
Eighteen-wheeler: Any semi-tractor/trailer.
Eights & other good numbers: Love and kisses, best regards, goodbye.
Eighty-eights: Love and kisses.
Evel Knievel: A motorcyclist.
Eyeball: Face-to-face.
Eye-ee 10-4: Apparently a gung-ho way of saying 10-4.
Eye in the sky: A police speeder-spotter aircraft.

Feed the bears: Pay a speeding ticket fine.
Feet: A linear amplifier.
Fifty dollar lane: Left-hand lane.
First personal: The operator's first name. (Mostly SSB use.)
Five by five: A signal that's strong with good copy.
Flip-flop: The return leg of a round trip.
Footwarmer: A linear amplifier.
Forty Roger: Somehow, it means 10-4, OK, fine, I agree!
Four: Short for 10-4.
Four ten: A cute way of saying 10-4.
Four wheeler: A passenger car, station wagon, van, or pickup truck.
Fox Charlie Charlie: The FCC.
Front door: 1st vehicle in a convoy, watching for smokies/radar.
Fuzzbuster: Trade name for a brand of radar detector.

Gas can: Service station.
Getting out: Being heard by other stations.
Give me some numbers: If you ask another station to give you a signal report, the operator may ask you to give him some numbers. This is because he thinks that his S-meter won't register properly if you just speak normally. You are then supposed to say "1-2-1-2-1-2" (no other numbers will do) and he will then tell you (in increments of "pounds") how many S-units your signal registered. Don't ask for me to explain this bizarre ritual, I suspect it goes back to the Druids and its rationale was lost along with the formula for turning lead into gold.
Go ahead: Invitation for another CB'er to transmit.
Go, breaker: The only known way to get someone to stop repeating "Break."

Go-juice: Gasoline or diesel fuel.
Go 10-100: Restroom stop.
Gone: Contact ended.
Goodbuddy: Once a term of address for any CB'er. Now the CB term for "gay."
Good numbers: Best regards, love and kisses, goodbye.
Got a copy.: Can you hear me?
Got your ears on?: Are you on the air?
Granny lane: The right-hand lane.
Greenstamps: Folding money, also speeding tickets.
Guy: A form of direct address to a CB'er, not especially friendly though.

Hammer: The accelerator pedal.
Hammer down: Accelerate to top speed.
Hammer lane: The left lane.
Handle: The operator/station name used for on-the-air identification.
Harvey Wallbanger: A speeding vehicle making sudden lane changes.
Hello Skipland: The AM-mode equivalent of CQ or CQ-DX.
H.F.: Unauthorized frequencies above Channel 40/27.405 MHz.
High roller: A speeding vehicle.
Hole in the wall: A tunnel.
Home 20: Where a CB'er lives.
Hook: A tow truck.
How am I hittin' you?: Request for a a signal strength report. See "Give me some numbers" for further instructions.
How's it look over your donkey?: Request for traffic/smokey info in the direction from which you just came.
How's it look over your shoulder?: (Ditto. Sometimes stated as "give me an over-the-shoulder.")

Indians: Television interference. (Also called "TVI" or "Tennessee Valley Indians.")

Jamboree: An organized CB gathering/convention/picnic.
Japanese junk: Term of endearment referring to a Yaesu FT-101 ham transceiver.

Keep the bugs off your glass & the bears off your ass: Drive safely.
Keep your nose between the ditches & smokey out of your britches: (Ditto.)

Keying up the mike: Pressing the mike's push-to-talk (transmit) button.
Kodak: A police radar unit.
Kojak with a kodak: A police radar unit.

Lid: A rotten operator-- rude, stupid, and without any regard for the most basic operating procedures. Mostly used by Sidebanders-- and never in a joking manner.
Linear: A linear amplifier.
Linear amplifier: An illegal transmitter-power booster.
Local yokel: A city/town/village police officer.
Lollipop: Astatic D-104 microphone.

Mail: Conversation(s) on a CB channel.
Making the trip: Signals getting through the interference.
Marker: Highway milemarker.
Meat wagon: Ambulance.
Mercy, mercy: Cornpone way of expressing happy surprise.
Mickey Mouse: Adjective used to describe anything dinky.
Milemarker: Highway distance post.
Mixmaster: Highway interchange cloverleaf.
Mobile: A CB rig installed in a vehicle.
Modulating: Talking on the CB.
Monitor: Listening to a CB channel.
Montfort lane: The left-hand lane.
Motion lotion: Gasoline or diesel fuel.
Motormouth: A CB'er who talks too much.

Needle in the sky: A tall radio broadcasting tower.
Negative: No.
Negative contact: Contact not made.
Negative copy: I can't hear the other station.
Negatory: No. (Also "negator.")

O.M.: Old man (all male CB'ers are O.M.'s, regardless of age.)
On the side: Standing by for calls. (Mostly SSB use.)
One time: Request for a very brief CB contact.
Over: Invitation for another station to transmit.

Pavement princess: A hooker operating from a motorhome parked at a highway rest area for the purposes of accomodating passing truckers and motorists. Many have CB rigs and make themselves known on Channel 19.
Peanut butter in your ears: Your receiver is no good.
Pedal to the metal: Accelerate to full speed, all clear ahead.
Persuader: A linear amplifier.
Pickum-up: Pickup truck.
Picture taker: Police radar
Picture window: A location on the highway where you can get a quick passing look at the traffic flow on another road you'll soon be entering.
Piggy bank: Toll booth.
Pit stop: Fuel/restroom stop.
Plain wrapper: Unmarked police car, sometimes the color is included (plain white wrapper, plain blue wrapper, etc.).
Polaroid: Police radar unit.
Portable: A mobile installation. (Mostly SSB use.)
Portable chicken coop: Movable truck weighing station.
Pository: Cornpone way of saying OK, yes, fine, I agree 100%.
Pounds: S-meter units (i.e., 9 pounds = S-9 on the meter).

For reasons not fully understood, no matter where you hail from, you're sort of expected to pronounce certain words and phrases a specific way-- with a style somewhere between Atlanta and Arkansas. All part of the mystique of CB radio!

Power mike: A mike with a built-in preamp to boost effective communications.
Preamp: Short for preamplifier, a device to boost audio or signals.
Pregnant roller skate: Any Volkswagen, especially the Beetle.
Puddinhead: A CB'er whose total conversation is a summary of what you said.
Pull the big switch: To leave the air.
Put an eyeball on: Take a look at.
Put the good numbers on you: Wish you best regards and end the contact.
Put the hammer down: Accelerate to top speed.
Put the peddal to the metal: (Ditto.)
Put Your boots on: Turn on your linear amplifier, I can't copy you.
Put your shoes on: (Ditto.)

Q-Bird: A device that produces a warbling 2-tone sound to attract attention.
Quickie break: Asking to use the channel for only a few seconds.
QSL card: A souvenir card from a CB'er with his handle and/or sideband ID number, address, favorite channels, and perhaps an illustration.

Radar Alley: Interstate 80 throughout Ohio.
Radidio: A CB transceiver (supposedly humorous).
Radio: A CB transceiver.
Radio Check: A request for a report on whether or not someone's CB set is working properly. Generally used as a ploy to initiate a brief, superficial, and random contact by CB'ers who really have nothing to say.
Radio shack: The room of the house wherein the CB set is used.
Ratchetjaw: To talk on CB, or someone who talks too much.
Ratpack: A bunchup of vehicles moving on a highway.
Read: To copy or understand a transmission.
Read the mail: To listen to the CB conversations without joining in.
Restum-up: An area along a highway designated for vehicles to pull over.
Rig: A CB transceiver.
Ringing your bell: Calling you via CB.
Rock: A plug-in transceiver device in older CB sets, used for determining freq.
Rocking chair: A vehicle located mid-point in a convoy.
Roger: Yes, OK, I understand.
Roger D: (Ditto, short for "Roger dodger.")

Rollerskate: A small car, especially a VW Beetle.
Rolling roadblock: A vehicle traveling far below the posted speed limit.
Rubberbanding: Stop-and-go, bumper-to-bumper traffic on a highway.
Rubberduckie: A short, rubberized, whip antenna, especially for a handheld rig.
Running barefoot: Not using a linear amplifier.
RV: A 4X4 vehicle.

Sandbagging: Monitoring CB conversations, but not joining in.
Sandwich lane: The center lane.
Scatterstick: An omni-directional base station antenna.
Seat cover: A woman in a vehicle.
Sevens: Best regards.
Seven three: (Ditto.)
Seventy thirds: (Ditto.)
Shack: The base station radio room.
Shake the bushes: Watch for police radar & advise others via CB if spotted.
Shoes: A linear amplifier.
Shoot skip: To communicate with CB'ers around the nation and world.
Short short: A very brief CB contact.
Shortie break: Using the channel for only a few seconds.
Shout: Calling or contacting another CB station.
Sideband numbers: Numbers used by sideband operators for ID instead of handles.
Sidebander: An operator who uses single-sideband mode and the special operating techniques and frequencies preferred for use in that mode.
Sidewinder: A Sidebander.
Single-sideband: A specialized transmission mode offering (legal) high power and a coverage range far exceeding that of AM CB stations.
Six-wheeler: A city-delivery truck, or car pulling a utility trailer.
Skate: A small car, especially a VW Beetle.
Skip: DX signals propagated by ionospheric conditions.
Skipland: All of the stations that are heard/worked via skip propagation.
Skating rink: A slippery road, especially due to icing.
Skyhook: A base station antenna.
Slab: A highway.
Slider: An unauthorized modification to a single-sideband transceiver that permits the clarifier control to vary the transmitter frequency.
S-Meter: A panel meter on a CB rig that shows the relative strength of incoming signals.
Smokey: Any county/state police officer, or a deputy sheriff.
Smokey break: Information regarding the location of a police car or radar trap.
Smokey with ears: A CB-equipped police car.
Splash: Causing interference on adjacent channels due to overmodulation or a linear amplifier.
Splashover: (Ditto.)
Speak to me!: A plea for a station to respond quickly to a transmission.
Stack them eights: 88's, meaning "love and kisses."
Stack them numbers: (Ditto.)
Stepped on: A CB transmission that was zapped by another operator.
Stick: An omni-directional base station antenna.
Sucker: A CB rig laid up for repairs.
Superslab: An Interstate highway.
Supertrooper: A over-zealous police officer giving out tickets for minor infractions.
Sweep the leaves: The "back door" vehicle in a convoy watching for smokies.

Take it to (channel #): I'll meet you on Channel (#).
Taking pictures: A police radar trap.
Talking skip: Working distant stations received via skip.
Talking snake: Any CB rig made by Cobra, especially the Cobra 2000GTL.
Talk Power: The ability of a rig's audio & signal to reach out.

Ten pounder: An especially strong signal.
Ten roger: OK, fine, I agree!
Threes & Eights: 73's & 88's, meaning best regards, love & kisses.
Threes upon you: Wishing you best regards.
Throwing a carrier: Jamming a channel with a dead carrier (also "chucking").
Ticks: Minutes, 60 to each hour.
Tie a ribbon on it: Last transmission in a contact before signing off.
Tijuana taxi: A marked police car, especially with roof lights on.
Town: A suffix CB'ers often add to any location name for apparent humorous effect, as in Chicagotown, Miamitown, Dallastown, Seattletown, etc.
Trip: The ability of a signal to make it through a crowded channel, as in, "How am I making the trip?" meaning "Can you copy me through the other stations?"
Truck 'em easy: Wishing trucker a safe journey.
Trucker: One who drives an 18-wheeler-- <u>never</u> a "truck driver."
Truck stop commando: A CB'er infatuated with talking to and hanging out with truckers at truck stops-- usually dressed in a fancy embroidered western-style satin shirt with sterling collar points, a quilted vest, 10-gallon hat with rainbow featers in the band, large ornate belt buckle, and silver-tipped cowboy boots-- mostly drives a VW Beetle with twin mini-whip antennas.
Twenty: 10-20, a location, as in "the home twenty."
Twin truckers: Dual phased mobile antennas mounted on outrigger side-view mirrors, as used on 18-wheeler installations.
Twisted pair: The telephone. (Also the "landline.")
Twister: A highway interchange.
Two wheeler: A motorcycle.

Uncle Charlie: The FCC.
Unit: A CB rig, or base or mobile station.
Unlocked clarifier: Same as a "slider" (q.v.).
Up one (or Down one): Meet you on the next channel above or below this one.
Upstairs frequencies: Unauthorized frequencies above Channel 40.

Walk all over: Communications knocked out by strong-signal interference.
Walk it on back: Invitation to reply to a CB transmission.
Walk on: To destroy a transmission with a stronger signal.
Wallbanger: A high speed vehicle making sudden lane changes.
Wallpaper: CB QSL cards.
Wall-to-wall: Very strong, also means filled to overflowing.
Wall-to-wall, treetop tall: Very strong CB signal.
Wall-to-wall, 10 feet tall: A signal so strong that it pins the S-meter on maximum.
Wall-to-wall bears: Police cars are everywhere.
We down: The contact is ended.
We gone: (Ditto.)
We're down & on the side: The contact is over, & new calls are invited.
We're up, we're down, we're clear, but we're around: (Ditto.)
We're tryin': Despite interference, I hope you can hear me.
What's your twenty?: Where are you located?
Willy Weaver: A drunk driver weaving all over the highway.
Worked on: A CB rig that's been modified for unauthorized extra frequencies.
Wrapper: The color of a vehicle, especially a police car.
W.T.: A walkie-talkie.

X-ray machine: Police radar unit.
X.Y.L.: The wife of a CB'er.

Y.L.: A "young lady." All women may be called Y.L.'s, regardless of age.
Your 'phone is ringing: Advising a CB'er that someone on channel is calling.

Zipper: Zip code.

CB 10-Code & SSB Conversions

The CB 10-Code is different than the official APCO Police 10-Code, although it was created in the 1960's in imitation of the APCO 10-Code. You may hardly ever hear some of these codes actually used, but we have included them all anyway. These codes are encountered only on the AM channels, Sidebanders never use them. For the convenience of AM operators converting to Sidebanding, the approximate (or closest) sideband equivalents in the Q-Code are shown in parenthesis.

A 10-Code may be used as a statement, as "10-4" meaning "OK, message received." It can also be stated as a question, as "10-4?" That means, "Did you receive my message." More commonly used 10-Codes are shown here in **boldface** type.

10-1 Receiving you poorly.
10-2 Receiving you well.
10-3 Stop transmitting. (QRT)
10-4 OK, message received.
10-5 Relay message. (QSP)
10-6 Stand by, busy. (QRX)
10-7 Leaving the air. (QRT)
10-8 In service. (QSK or QSX)
10-9 Repeat message.
10-10 Standing by. (QSX)

10-11 Speak slower.

10-12 Visitors present.

10-13 Advise road/traffic conditions.

10-16 Make pickup at: _____.

10-17 Urgent.

10-18 Any messages for me?

10-19 No messages, return to base.

10-20 My location is. (QTH)

10-21 Call me by 'phone.

10-22 Come in person to: ______.

10-23 Stand by. (QRX)

10-24 Assignment completed.

10-25 Contact: _____.

10-26 Ignore last information.

10-27 Changing channel to:__. (QSY)

10-28 What's your ID? (QRZ)

10-29 We've talked too long.

10-30 Unauthorized operation.

10-32 Radio check.

10-33 Emergency.

10-34 Trouble here.

10-35 Confidential information.

10-36 Correct time.

10-37 Send a tow truck.

10-38 Send an ambulance.

10-39 Your message delivered.

10-41 Change to channel:___. (QSY)

10-42 Traffic accident.

10-43 Traffic tie-up.

10-44 I have a message for you.

10-45 Calling all stations. (QSK or CQ)

10-46 Assist motorist.

10-50 Break channel (QSK or "On the side")

10-60 Send your next message.

10-62 Cant copy you, call by 'phone.

10-63 Network directed to:_____.

10-64 Network clear.

10-65 Awaiting assignment.

10-67 All units, comply.

10-69 Message received.

10-70 Fire at:_____.

10-71 Send your traffic.

10-73 Speed trap at:____.

10-74 No.

10-75 You are interferring. (QRM)

10-77 Negative contact.

10-81 Reserve motel room for:___.

10-84 Call 'phone number: _____.

10-85 Address is:_____.

10-89 Communications tech needed at:_____.

10-90 I have TV interference.

10-91 Speak closer to the mike.

10-92 Your transmitter is malfunctioning.

10-93 Am I on frequency?

10-94 Count from 1 to 10 slowly.

10-95 Transmit dead carrier for 5 seconds.

10-99 All units secure.

Truckers' 12-Code

Truckers once concocted their own private spoof 12-Code in an effort to needle CB'ers in passenger vehicles on the Interstates who had, in the mid-1970's, suddenly invaded the truckers' channel, Channel 19. Truckers looked with considerable annoyance if the code was used by non-truckers. Some old time truckers still use it from time-to-time.

12-1 There's one in every crowd.
12-2 Get your own towel, mine is already wet.
12-3 I can't believe I ate the whole thing.
12-23 If that guy transmits once more, I'm selling my CB gear.
12-24 A pill a day keeps the doctor away.
12-25 Shut up, stop ratchet-jawing on this channel.
12-35 Ride on...ride on into the sunset, heroes of the road!
12-37 Don't jive me, turkey.
12-38 Get off my back.
12-39 Beats the hell out of me.
12-45 So what, big deal!
12-46 Here we are gang, Miss America.
12-47 Beautiful, simply beautiful.
12-48 This is just not my day.
12-49 So much for you and the horse you rode in on.
12-50 Whatever it is, if you got it-- a truck brought it.
12-61 Don't bull me.
12-62 Go fight city hall.
12-65 Screwed up like Hogan's goat.
12-69 Excuse please; couldn't find my way.
12-72 You must have me confused with someone who gives a damn.
12-73 A fantastic screw-up is now in progress.
12-74 It figures!
12-75 Here I am-- Mr. Wonderful.
12-76 SNAFU
12-77 Hold on, I'll call the chaplain.
12-78 Oh my God, now he thinks he's a Smokey!
12-83 Who, me?
12-84 Can't receive you, there's bird crap on my antenna.
12-85 Let's wait, maybe someone else will do it.
12-86 I agree, you do have problems, dummy.
12-94 Hey buddy, you've got crankcase oil for brains.
12-96 CB maniacs of America, unite!
12-100 Just one more cottonpickin' trucker!
12-200 Don't mess with me, pal.

CB'ers 13-Code

Another irreverent CB code originated in 1976. Called the 13-Code, it seemed to sum up so many things not covered by the 10-Codes:

13-1 All units copy you and think you're an idiot.
13-2 Yes, I copy you but I'm trying to ignore you.
13-3 You're beautiful when you're angry, big fella.
13-4 Sorry 'bout that, guy.
13-5 Same to you, Mack!
13-6 OK, so I goofed-- none of us are human.
13-7 If you can't copy me, the problem's at your end 'cause I'm running 3 kW.
13-20 Is that your mike clinking or are your uppers coming loose again?
13-21 My god, are you getting paid by the word?

13-22 Lady, is that your voice or did you install a steam whistle?
13-23 If you had yakked for another 30 seconds you'd have been eligible for a broadcast station license.
13-24 Y'know, you made more sense last time when you were stoned.
13-25 A bunch of us local operators chipped in to buy your rig from you. Have you ever considered stamp collecting?
13-26 Next time you eat garlic, would you talk further from the mike?
13-40 Yeah, your signal sounds great. Now shut off the rig and give me a landline so I can find out what you want.
13-41 Either my receiver is out of alignment or you're on Channel 57.
13-42 Either my speaker cone is torn or you'd better try it again when you sober up a little.
13-43 That was a great 10-minute transmission, now try it with your mike connected.
13-44 Your new rig sounds great. No wonder they discontinued that model.
13-45 Your transmitter must have a short 'cause there was smoke coming out of my rig during your last transmission.
13-46 That transmission was on your new antenna? You'd get a better signal out of a 6-inch piece of damp string.
13-47 What a fantastic signal! Give me a few minutes to bring my mobile unit into your driveway so I can get a copy on you.

The Channelmaster regards himself as a benevolent despot and treats his many adoring subjects as if they were so many small children vying for his favor. I once heard one of these guys seriously suggest to the users of his chosen channel that they give him a public service plaque. They didn't, and a few weeks later he abandoned their channel and took his good services to another frequency!

CITIZENS BAND RADIO STATION

4

How Far Can You Talk?

The FCC regulations specify that CB stations may communicate over distances not exceeding 250 kM (155.3 miles). This probably exceeds the groundwave coverage of any station meeting FCC technical standards. Any stations whose signals reach you from more than this distance are being propagated by ionospheric skip.

Groundwave signal coverage is dependent upon several variables including the antenna type and height above the average surrounding terrain, transmission mode, whether base or mobile, interference from other stations, power line noise, weather conditions, obstructions in the signal path, skip conditions, and other factors. Mobile units generally don't have the communications range of base stations. Single sideband (SSB) stations invariably have a greater range than stations using AM transmission-- all other factors being equal.

While it would be wonderful to be able to proclaim (as some have done) that mobile-to-mobile range is between 5 and 15 miles, base-to-mobile is 10 to 25 miles, and communications between base stations averages 15 to

30 miles. Actually, you might use such figures as a rule-of-thumb for AM mode operations, enhancing those figures by 50% to 75% when using SSB mode. Still, there are so many variables involved that such distances can't always be assured.

Some of the variables aren't under the control of the operator-- things such as weather, signal path obstructions, etc. Other variables are under the control of the operator, and one could then assume that an SSB base station utilizing a high gain directional rotatable beam mounted at high as legally permitted (highest point 20 ft. above roof of the building it is mounted, or 60 ft. above the ground), you'll achieve maximum groundwave coverage.

Skipland

Drifting around in the sky above our planet are a bunch of interesting things including artificial satellites, UFO's, what's left of the ozone layer, all sorts of cosmic particles, magnetic fields, X-rays, Van Allen radiation belts, plus various layers of the ionosphere.

The ionospheric layers have been given the job by nature of being involved with long distance radio communications. These layers can reflect high frequency radio waves, however their ability to do so isn't uniform. It is affected by the time of day, the frequencies involved, the season of the year, and an overall eleven year cycle that is tied in with sunspot activity.

The ionosphere therefore has a profound effect upon CB communications since the band allocated for the CB service lies within the portion of the spectrum most affected by these layers (roughly 2 through 75 MHz). The FCC, of course, has regulations that put the kibosh on communicating with other CB stations whose signals arrive from distant points by means of being reflected by the ionosphere (skip signals, as they are called). Nevertheless, when conditions permit, CB'ers have long been in the habit of trying to "work DX," or "shoot skip" when there's a "band opening." Sometimes CB'ers are nailed by the FCC for such unauthorized practices but, especially when the

activity is helped along by overpowered equipment or else takes place outside the confines of the authorized band.

A popular tongue-in-cheek CB axiom of the "Skipland" fan says to "remember that it isn't legal to talk to a station 600 miles away-- but it's OK so long as you only talk 155 miles and the other station listens the rest of the way."

Whether you talk to skip stations, just listen to them, or cuss them out because the're bombing in all over your attempts at local communications, you can't escape their existence on 27 MHz.

The sunspot cycles that are major factors in determining DX propagation have come and gone several times during the life of the CB service. Scientists began counting sunspot cycles in the year 1720. The high point of Cycle 19, bringing with it once-in-a-lifetime DX propagation, began in the mid-1950's and ended in the mid-1960's. The FCC's opening of the 27 MHz band for CB purposes coincided with the best DX portion of this cycle.

Cycle 20 (nowhere near as good as the previous one) peaked in the late 1960's, Cycle 21 peaked out in 1979, the Cycle 22 peak is 1990, with DX conditions remaining good for the following couple of years.

When the sunspot count is high, the result is heavy ionization of our atmosphere. It is during those most active periods (at cycle peaks) that the higher frequencies can be reflected by the ionosphere. During times of low sunspot activity, perhaps the maximum usable frequency (MUF) will be about 15 MHz. The only signals to be heard on 27 MHz will be locals. Signals on frequencies above the MUF aren't reflected-- there's no skip propagation on CB.

As we enter each Cycle, sunspot activity increases and slowly the MUF starts to rise. During peak years, 27 MHz becomes richly endowed with signals from distant lands. Of course, it isn't a steady 24-hour-a-day thing going every day of the year. Moreover, there are temporary disturbances that take place from time to time that knock out DX propagation. One such event is a "sudden ionospheric disturbance" (SID) that can flash by in only a few

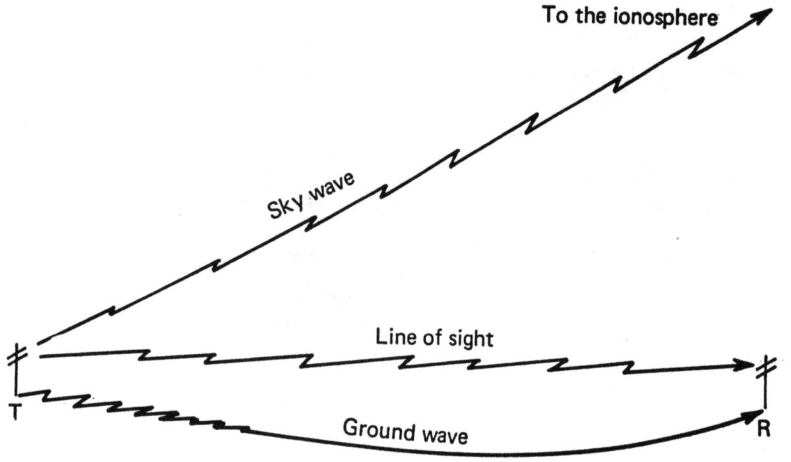

When a signal leaves an antenna it can travel to a receiver 3 ways-- along the ground (groundwave), directly from the transmitting antenna to the receiving antenna, and via ionosphere (sky wave). See the chapter on base station antennas for more information on groundwaves.

minutes or last for about an hour. SID's can be so severe as to smother all but the signals from stations located in your immediate vicinity.

Ionospheric storms, although not as dramatic as SID's, can weaken and cause flutter fading on signals for several days at a time, both day and night.

The term "skip" relates to the fact that some signals literally skip over the space between their point of origin and the point of reception. This phenomenon is the result of the fact that the signal doesn't follow the curvature of the earth's surface-- but zooms off in a relatively straight line up to one of the ionospheric layers and then bounces back down. A receiving station over the horizon and out of groundwave range of the transmitting station wouldn't hear the signal. Another station hundreds or thousands of miles away-- where the signal happened to come back down-- would receive it loud and clear. The area in between (where signals aren't heard) is known as the skip zone because the signals skip over it.

In looking at the diagram, note that the lower the angle of radiation, the greater the distance from the point of origin the signal will be when it returns to earth. The geometric law that the angle of reflection equals the angle of incidence is not absolutely true. Apparently some bending of the radio wave occurs, rather than a point-to-point reflection. But for most purposes we can assume that the radio wave will be reflected or refracted back to earth at nearly the same angle as it arrived.

At first it was thought that there was only one layer up there that was doing all of the honors in bouncing the signals. It was known as the Kennelly-Heaviside layer in recognition of its discoverers. Eventually it was realized that there were several ionospheric layers and they varied in their effective height above the earth.

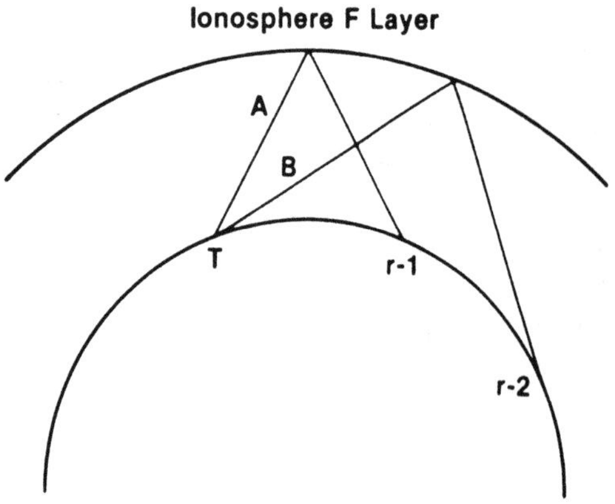

Showing that high angle signal at A will travel less distance than the low angle signal at B.

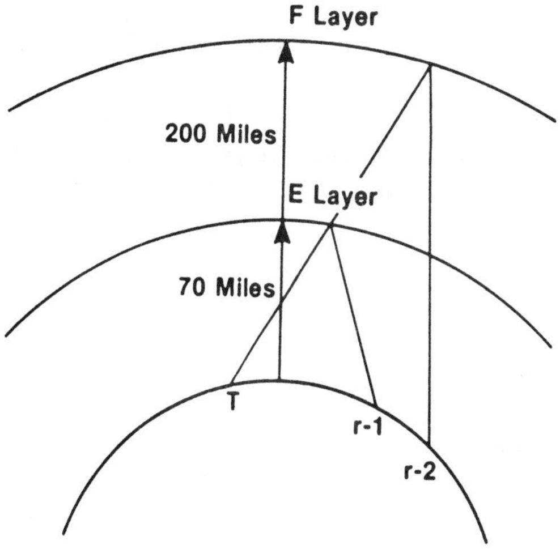

Showing radio wave reflection off of both E and F layers. Distance at r-2 will be twice that at r-1.

The diagram shows the layers concerned with long distance propagation. The E layer is about 70 miles up. The F layer is actually two layers, F-1 at about 140 miles altitude, and F-2 at around 200 miles. At night both F layers join into a single layer about 175 miles high. Not shown is the D layer which exists only around noon and affects frequencies below 5 MHz.

Besides these layers, there are also transitory ionized regions that show up on a sporadic or irregular basis. These generally occur at a height of about 60 to 70 miles and they are known as Sporadic-E. The phenomenon shows up most frequently during the summer and during the hours of daylight-- and sometimes during winter nights. The DX benefits such a phenomenon brings may last for several hours and affect only a limited geographic area which travels towards the west. Since Sporadic-E normally produces skip at 27 MHz, its long been known as one of the more important factors in DX propagation for CB'ers.

It's important to understand that none of these layers are solid walls that bounce signals like ping-pong balls. They are ionized areas that vary in thickness. They are constantly changing in composition and the degree of ionization at a given time determines if the layer will absorb most of the radio energy, reflect most of it back, or allow the energy to pass right through into outer space.

The amount of ionization is dependent upon the sun's level of activity in generating solar energy in the form of ultraviolet radiations. This is related to sunspot activity-- sunspots being isolated cooler regions on the solar surface. These sunspots are sources of enormous amounts of energy that pour in a stream of hydrogen and calcium gasses in a cyclone-like vortex between the two poles of an apparent magnetic loop.

The phenomenon of reflection is very dependent not only on the degree of ionization, but also on the frequency of the radio wave. And that later is a variable since the frequency that will be bounced back is constantly changing in direct relation to the ionization. Less ionization is required to reflect the lower frequencies than the higher frequencies.

At any one time, for a particular zone of the layer, there will be a MUF-- the highest frequency that will be reflected from a specific area of the layer.

The layer isn't a homogeneous mass with uniform thickness and density. Rather it is constantly expanding and contracting, moving up and down, and varying in its amount of ionization. For any particular signal path between two distant points on earth the layer will be in a given state. It will reflect signals below a certain maximum frequency. Above that frequency, DX will not be possible.

The MUF, therfore, isn't constant each hour, but must be calculated over a specific radio path, taking into account the degree of ionization that will exist over that path. In general, you'll get a pretty good idea of possible radio paths if you can visualize the illumination of the earth as if you were looking at it from outer space. The TV pictures our astronauts sent back of the terminator line (the edge of darkness and light) from the moon were spectacular proof that half the planet is in relative darkness half of the time. There's a wide variation depending upon time of year and latitude, but while the rays of the sun are hitting, the layers will be ionized to some degree.

Radio signals will travel over the path that offers the best reflecting surface back to earth, it makes no difference whether or not you intend for them to do so. If you are using a ground plane, a coaxial antenna, a whip, a 5/8-wave vertical, or other omni-directional antenna, you pretty much have to take what comes. You'll notice that the radio path generally follows the path of the sun. As signals fade from the east, they improve from the west. This is directly attributable to the decreasing ionization as the F layer thins out and splits into two separate layers.

Generally the first frequencies to lose the ability to reflect off the ionosphere are the higher ones, since they require more ionization. First CB and the 28 MHz ham bands will drop out of the DX parade, then, later the 21 MHz ham band will feel the effects. During the winter months, even the 14 MHz ham band will lose its DX abilities and skip propagation will work well only on the 7 MHz and lower frequency ham bands.

During peak periods of sunspot cycles, the 14 MHz ham band will be open for DX to some spot on earth during summertime on a round-the-clock basis, the 21 MHz ham band will often be open almost till midnight, and even CB and the 28 MHz ham bands may produce DX well into the evening.

Now, before you write to tell me about hearing CB skip roll in late at night, let me again remind you about the Sporadic-E type of DX previously mentioned. This phenomenon provides 1-bounce DX over a 1,250 mile maximum path. Those very loud DX stations you hear coming at you from out of the northwest are invariably the result of Sporadic-E.

Planned DX is really possible only with the use of a directive antenna, a rotary beam. In order to make maximum use of the optimum radio path from a given starting point and that of a distant station, the operator must know something of both the probable area of best ionization and the correct heading in which to point the beam antenna.

Most operators assume that means pointing the beam right at the other station. That's correct, but the question then becomes to define the meaning of "right at." If a person is used to looking at the typical mercator projection map, they are going to have to unlearn the idea of directions when it comes to correctly pointing a beam antenna.

From the time of childhood, we are shown maps that seem to indicate that certain areas of the world lie in well-defined directions from our country. You want to know the direction of Rome from Chicago? Heck, they're almost on the same latitude. You can just about draw a straight line going east and connect the two cities. A beam antenna just needs to be pointed due east at 90 degrees to hit Rome from Chicago, right? Wrong!

Radio signals generally prefer to take the shortest path between two points. Yes, that's a straight line. But it's using the "great circle" path--the same route international airliners take when they want to go non-stop the fastest way. You can get a good idea of this path if you pin a string on your home town as it appears on a globe of the world then place the other end of the string on Tokyo, Melbourne, or Stockholm. That represents the shortest path.

If you don't have a globe handy, use an orange and visualize where the nations are. Now you will find that a heading from Chicago to Rome is actually northeast, not east as you expected.

Sky wave signals are reflected by the E or F layers. Usually, summertime DX results from the E-layer. During periods of high sunspot activity, the F-layer can skip CB signals over long distances. Sometimes 2-hop propagation is possible via one of the layers of the ionosphere. In that case, signals travel up to twice normal ranges. Sometimes distances are even greater than this!

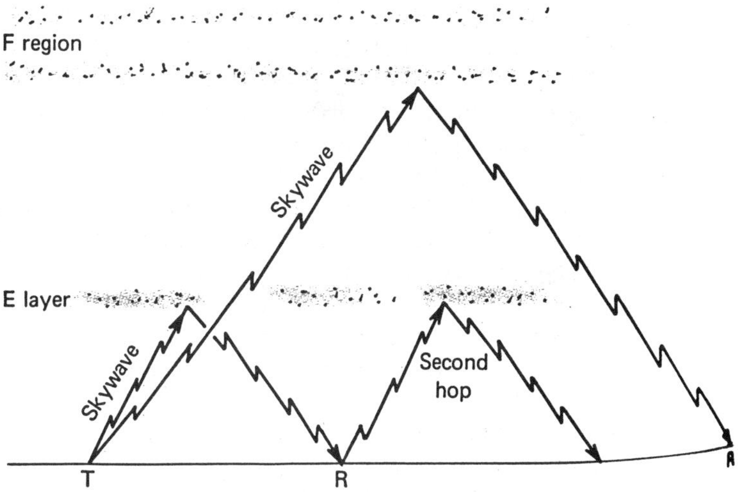

With some practice, you'll quickly start to think in these new dimensions and directions. You'll realize that aiming over the polar regions is often the shortest path. A great circle chart of the world, with the center near your location, will aid you in pinpointing the optimum beam heading.

Even a shot from Los Angeles to New York would be east-northeast about 60 degrees, despite the fact that that we would normally think of New York being due east. You should learn to visualize the path a signal takes, where it hits the layer, where it will come down and bounce again. Yes, I said bounce up again!

Even with a low angle of radiation and a bounce off the F layer, the longest single hop distance is around 2,500 miles. The distance from San Francisco to Rome is 6,228 miles via the shortest (great circle) path. This can't be a one-hop trip-- it would take three or four hops, depending upon the best path and the angle of radiation.

This isn't done for free, however. Each hop probably weakens the signal by about 10 dB. That means only one tenth of the power gets back on up for the next hop. You can figure what a multi-hop path will cost in the way of lost signal strength. Even with that, you'll hear signals from half way around the world bombing in at 30 dB over S9. Don't think that signals aren't heard because they have to make several hops.

On The Beam

Let's see how all of this works in actual practice by seeing how signals go from one place to another and keep their moxie. Let's assume that there is a station in Los Angeles. On a mercator projection map, it's seen that the city is about 34 degrees latitude and 118 degrees longitude.

The operator in Los Angeles has a beam antenna that concentrates his signal in a particular direction. The first thing he wants to do is calibrate the antenna direction indicator so that it shows, in degrees, the direction the in which the antenna itself is pointing. To do this, he has to know the deviation between magnetic north and true north, since all all bearings are based on true north. For Los Angeles, the variation is 17 degrees, and that correction was taken into account before the boom-to-mast coupler was bolted into position. Navigational charts used by aviation and maritime interests invariably indicate the magnetic deviation information.

It helps to keep track of DX data with a 24-hour station clock set for Coordinated Universal Time (UTC-- formerly and sometimes still called GMT). This provides information on the local time in most parts of the world-- data that's of use in calculating how much of the signal path between two stations is, or has recently been, in sunshine.

One side of a great circle pointer. The bearing of 197° would put the signal into Israel via "long path."

With a great circle chart having a movable pointer, the optimum beam heading can be determined for any given location. For the shortest path from Los Angeles to Germany, the bearing shows up as being 30 degrees. That means that the signal would have to bounce down somewhere in Canada's Hudson Bay, take off again over Greenland and come down in the snow, and go up again. It's a 5,600 mile haul so the third hop brings the signal right into Germany. During a 24-hour period there are two likely times that produce the best chance for such a trip in the 14 MHz ham band. One is early morning (when it is late afternoon in Germany). The other is when it is late afternoon in Los Angeles and morning in Germany. Bands as high as 21 mHz and even CB and 28 MHz are likely to be open at those times.

You may have realized that stations on the East Coast have some advantage when it comes to European skip compared to the West Coast. Their signals don't pay the 10 bB penalty of an extra bounce. But West Coast stations get the same bargain with signals to Japan and Australia.

Long Way, Wrong Way?

There are times when the best route between two locations is the so-called long path. It's called that because radio signals are bouncing the long way around the earth. Again, it depends upon the state of ionization existing between the stations. The longer path may actually be the stronger path.

When you hear a peculiar echo effect on a DX signal, you know that it's likely that you're hearing the other station's signal from both directions at almost the same time. That would be the time to spin the antenna around to face into the direction opposite the one normally used.

Radio propagation is a very complex topic and our capsule discussion here has barely even scratched the surface. It was intended to provide some general information on the nature of the beast. There are a number of excellent books available that contain highly detailed information on matters relating to propagation. You might wish to check out, for instance, The Shortwave Propagation Handbook, by George Jacobs (W3ASK), and Theodore Cohen (N4XX), from CQ Publishing, also various publications of the ARRL, including the Radio Amateurs' Handbook.

A look through the pages of ham periodicals such as CQ and QST will reveal ads from commercial sellers of great circle charts and also computer printouts and programs that show optimum beam headings centered on any desired location.

5

In Case of Emergency

Besides the joys of hobby use, and the possibilities for business applications, one of the most utilitarian potentials for CB is in its application for roadside emergency use. Channel 9 has been designated by the FCC for such purposes-- although there's no denying that there have always been abuses of Channel 9 that cut into its integrity. Moreover, in other nations of the world that have 27.065 MHz (Channel 9 in the U.S. and Canada) allocated for CB purposes-- it is usually not designated for limited or restricted usage. Therefore, when the band opens with skip, all sorts of distant stations roll in and can render the channel almost impossible to use except for very close-in comms.

Nevertheless, Channel 9 endures and constantly proves its value. That's because the toughest roots of the nation's strength lie in the heart of its community life. From our largest cities to our tiniest crossroad hamlets, the voluntary actions of concerned and responsible individuals and volunteer groups and teams, working in the public interest, characterizes one of the great blessings of an independent people-- the ability to help one's neighbor.

Thanks to this selfless willingness to help, and the availability of CB radio, a vital network of your neighbors is ready to be able to help out with instant handling of critical messages when help is needed...on the highways, in the farm fields, at camping sites, on the lake, at public catherings, picnics...almost anywhere accident, disaster, sudden illness, or even a lesser emergency occurs.

The largest, oldest, and most well organized of emergency networks dedicated monitoring Channel 9 for emergency requests is REACT International (headquartered at 242 Cleveland, Wichita, KS 67214). Started in 1962, the initials REACT stand for "Radio Emergency Associated Citizens Teams." The organization

Channel 9 monitoring groups come and go. REACT is the only one heard from in recent years.

Channel 9 signs along the highways advise motorists that someone's there to help when needed.

is manifested through the participation of numerous local "teams." Each of these volunteer teams is wholly independent. Its administration and control is worked out to suit local needs and conditions by its own membership. Each team maintains a Channel 9 monitor station, standing by on Channel 9, often on a continuous basis. The chartered teams are located throughout North America and in many overseas areas. The national organization offers a framework for the teams to work within, an excellent publication called The REACTer, plenty of advice, coordination, and ideas on emergency operations. REACT's distinctive "Monitor 9" road signs let motorists know that they're in an area where there's an active REACT team.

While REACT's volunteers often receive media publicity for their roadside free coffee/doughnut roadstops during heavily-traffic holiday weekends, the majority of the time their work is unsung. Helping to change tires and bringing

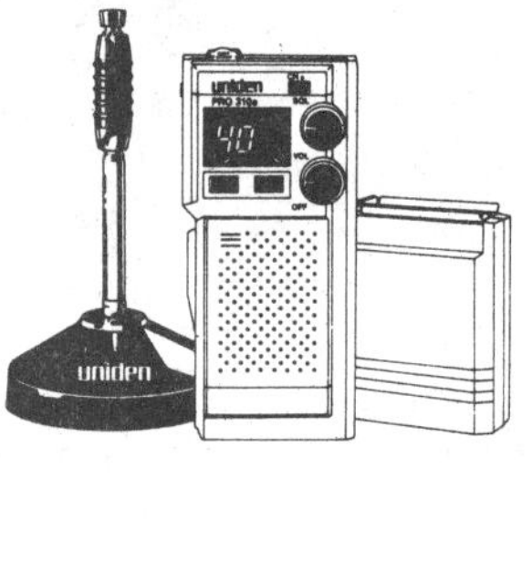

CB's emergency advantages for those who don't want to be CB'ers comes in the form of hand-held 40-channel AM transceivers that come with mag-mount antennas. You keep them in the vehicle's trunk or glovebox until they are needed, then plug them into the car's cigarette lighter for power and shout for help. Many companies now make these inexpensive units-- the one shown here is the Uniden PRO-310e. Weighs only 3 lbs. Mostly, this type of rig runs less than a hundred dollars. They're handy!

New York State Escapee Caught; Boy Gets Credit

An escapee from the Ulster County Jail in Kingston was nabbed by sheriff's deputies but authorities gave credit for the arrest to a 15-year-old boy using a citizen's band radio.

"It was actu...

BLEEDING ARIZONA YOUTH IS RESCUED

A youth who slashed his wrist in a hunting accident was saved from serious injury by the combined efforts of citizens band radio operators and a private pilot.

Miles Swain, 13 ...

Philadelphia, Pa. CB Operators Track Down A Stolen Car

Clarence Thompson heard a real-life drama on his Citizens Band radio.

He listened as four of his CB radio buddies closed in on the man suspected of stealing his car.

As Thompson sat by his h... home, voices crackled ...

"I see hi...

another CB operator repo... ted the car around ...

The man said ...

CB CLUB Gives $1,880

Barking Water (Okla.) CB Club members voted to give a total of $1,880 away to "good causes," including ... Davis Fund.

... club to help build ... between ... Roller ... 56. ... Children's ... from ... ceive $30 ... their ... Coaste ... Mi ... Hor ... th...

Truckers, CBs Foil Robbers

CB radios and two truck drivers were the heroes after a West Memphis, Ark., woman was assaulted by two men attempting to rob her at the rest area on Interstate 55 north of Fruitland.

But before it was all over, gunfire was exchanged at the rest area, injuring one ... the assailants.

"... really saved us," said truck ... Lively, who assisted the ... her car keys from ... to block

door, her face bloody, he said she stood behind him and he demanded the men return her car keys. They returned the keys and walked toward their car.

Another truck driver, Philip W. Brough of Raleigh, Tenn., came to assist. He was given a .32-caliber gun by Mrs. Stewart that she had in her car.

At least nine shots were exchanged.

A Cape County sheriff's deputy, Lt. Warren Rauh, said four shell casings from a .22 rifle were found where the assailants' car was parked and five casings from Mrs. Stewart's gun were found by a picnic table that Mr. Brough sought cover behind.

One of the assailants, Roscoe Pittman, 27, of Chicago, Ill., was wounded in the head during the shooting, Lt. Rauh said.

The two drove around Mr. Lively's truck, which was blocking the entrance to the interstate, but were apprehended after exiting south on Highway 61 by Cape County sheriff's officers and taken into custody.

... Stewart was taken to St. Francis ... treated for bruises on ... right eye, and

Connecticut CBers Credited In Boat Rescue

A group of citizens band radio operators helped police rescue a stranded boater and his 15-year-old son after the two had engine trouble on Long Island Sound.

Robert Salfton of Hamden, credited the CB operators with their rescue. "In a green boat, the same color as the water, they never would have found us without the help of the CBers ..."

Police heard the first ... layed from Madi... tenden. ...

Vt. CB'ers Help A Youngster

A 400-vehicle convoy raised $800 t... $1,000 for a disease-stricken boy Shelburne Falls whose leg was am... tated at the hip. The CB folks also ... plied him Tommy or Wonder T... ... rig. ...

San Diego CB'ers Aid Police/Sheriff

A crime alert program aimed at enlisting CB operators was put into effect by Sheriff John Duffy and ... Bill Kolend...

The Citize... all CB opera... port suspiciou... channel 9 to ... relay the info... ment agencies ...

"It's pretty ... and myself th... ough law enf... d. "We need ... n that inclu... dded Kolen... d of the pu... e than we ...

Lady No Name Directs Rescue Efforts By CB In Missouri

Without the help of Lady No Name, Guppy, Cartoon, Starship, Mouse and others Audrey Felts might have become a traffic fatality victim.

Miss Relts, 47, was driving ... plunged ...

But, Miss Felts had a CB set managed to call for help. Her di... signal was heard by Lula Farmer ... sas City, Mo. ...

Dozens Hear 'He's Shooting At Me' As Ontario CB Radio Fan Helps Nab Raiders

Dozens of ... drama on the ... as a teenage ... at high ... He's sh... ne," ... his ... us ...

...ed to the ... radios ... ob-

read the license plate, and we were going very fast.

"Finally, when Moonbeam Base believed me, all the other users got off ... air and Moonbeam phoned the police and told them what was happening. ... guys in the car ahead realized ... following them and the passen... window and fired his ... luckily he missed,"

... the dash- ... shotgun

CB "Skip" Saves Lives!

A man and a boy stranded in Ontario in —40c weather last December scribbled their names on the windshield of the vehicle to identify their bodies for rescuers after they had abandoned any ...

Const. Wroblowsky estimated the two would have ... three or four hours had M... OPP as to th...

After rece... Kenora OPP who then search and ...

The latte... the Borde... club.

OPP m... and the C... tou Acce...

never ... 3, of ... The ... iette, ... gging ... bout

CB Frees Texas Rancher Tied Up By 2 Men

Ranch foreman Travis Parker's experience could have been right out of an old Western movie script, except for one important modern element.

Parker was patroling the Bois D'Arc ranch south of San Angelo looking for lawbreakers (illegal hunters).

When he confronted two of the bad guys, one got behind him and knocked him out. They chained his hands to the steering wheel of his pickup, and tied his head back with baling wire.

And there they left him, to spend the night in that uncomfortable position and in the biting cold. The temperature in San Angelo was to dip exactly to the freezing mark be... the sun rose.

But Parker ... have to spend the night there ... companion was ... able to ge... medicament to ... the out... his res... T... ...ble ...

tunately had been left on.

"I was afraid I would knock it off," he said.

His voice strained by the baling-wire gag, Parker was able to contact a CB dispatcher. The dispatcher said later she thought the man sounded like he had suffered a heart attack.

His location was pinpointed, and Parker was freed. A bump on the head, and a few bruises, including his tongue, were the only ill-effects of his ordeal.

One change was brought about by the incident. Tom Green County ranchers now patrol in pairs, looking for "road hunters."

(A road hunter is a trespasser, since he hunts on land he has not leased.)

The story does not yet have a typical old movie western ending: The bad guys haven't been caught and brought to justice at S9 press time. But if Parker has his way, they will be prosecuted to the full extent of the law, and justice will triumph ... just as the script calls fo...

Penn. CB'ers Aid In Saving Hunter's Life

A gallant wintertime effort by three Noxen men on an ambulance assi... ment with the help of local CB'ers be credited with saving the life ... out-of-state man, who was the ... of a hunting accident. An unc...

Denver CB'ers Credited With Saving Man's Life

A team effort by several CB operators, unacquainted with each other except by voice, apparently saved the l... of Godfrey Mead, 49.

Mead, whose CB tag is "Color Poor Bo... ...fered a heart attack a... ...ing his truck. 7:15 p... He ... "10-3 ... Bo ... erat ... wif ... M

Penna. CBers Help Nab Suspects

Three Pittsburgh men were arre... for burglary as the result ... ing by a CB oper... don police said.

CB's Help Promote Road Safety

The CB radio boom, which began three years ago with a truck driver protest strike, has spawned a cooperative citizens law enforcement network which promotes highway safety in 30 states.

In 1974 truck drivers were using their CB radios to warn each ... "smokey's" ...

CB Crimewatch: N.J. Community Says "OK!"

The Madison Police Department (Madison, N.J.) is another example of CB radio making believers of law enforcement officials.

Police Chief Harry Bartow decided what was good for him was good for the ... had a set installed in his per... back and

...week of action Bar... saved the life of a 54-... town man who suffered ... while driving on Main ... Madison.

...serby, realizing the man was in ..., called for help using CB. Madi... police picked up the call and were ... The heart attack

Although some may consider CB a toy, it's not only a great emergency aid, but also has proven itself in many safety and humanitarian applications. It's a headline maker!

Some present/past Channel 9 monitoring groups give courtesy cards to motorists they've aided.

gas to stranded motorists doesn't make headlines, neither does pulling stuck vehicles out of the mud or snowbanks. Still, not all of REACT's challenges are routine-- members search the woods for lost children, instruct hospitals to await the arrival of injured persons being transported-- in other words, handle the dozens of local emergencies and incidents that occur every day. Usually, REACT teams work in close cooperation with public safety agencies in their respective areas.

Let it be noted that during the CB boom of the late 1970's, REACT had what seems to many to have had far more active local teams than it does today. Of course, during those years there were more CB'ers to staff those teams. It was a mixed blessing since, in the flood of members that entered the group's membership rolls there was an odd assortment of folks. While many were the dedicated type of members, other were of questionable motives and of shallow commitment to REACT's motives and goals. Some were lazy, others pictured themselves as self-appointed police officers, or persons who had been given FCC stewardship over communications on Channel 9.

This sometimes resulted in two REACT teams in the same general area bickering over geographic territories, or over who had priority use of the channel during simultaneous emergencies. Somtimes, the monitor on duty at a REACT station would attempt to kick independent (non-REACT) emergency organizations, or emergency base stations operated by individuals, off Channel 9 under the (erroneous) notion that Channel 9 was under the control of REACT or designated for the exclusive use of the group. In actuality, all CB stations have equal rights to use Channel 9, no group or individual can claim any standing exclusive rights to use or control the channel-- although common sense would dictate that stations would make every effort not to cause deliberate interference to stations engaged in vital communications.

Today, most of those hassles seem to have passed although isolated problems still seem to pop up on occasion. The organization seems healthy and growing and continues to issue new charters for local area teams. Further information on REACT International, its services, team locations, membership requirements, etc., can be obtained directly from the organization's headquarters in Kansas.

Those requiring the services of REACT or other teams or individuals on CB should tune to Channel 9 (AM mode) and ask for a base station. You may

Although having fewer teams now than during CB's 1970's boom years, the presently active REACT monitoring groups seem more courteous and efficient than ever.

wish to specifically ask for a REACT monitor. Some current mobile CB rigs are equipped with an "instant" Channel 9 button that switches the rig there with a minumum of fumbling.

Another excellent emergency CB aid is intended for those motorists who really aren't interested in having the joys of CB with them at all times, but would like to know that it might still be relied upon during any emergency. Therefore, many manufacturers now offer little, inexpensive, all-channel CB rigs that aren't intended for permanent installation. The full-power transceiver plugs into the vehicle's cigarette lighter, with a magnetic mount antenna that can be temporarily stuck to any metal part of the vehicle's body body. These rigs can be tossed into the vehicle's trunk or glovebox until such time as they are needed for summoning aid.

Many U.S. Coast Guard shore stations also monitor Channel 9, although the preferred way of contacting the Coast Guard is on VHF-FM Channel 16 (156.80 MHz). The agency might not provide CB Channel 9 monitoring services at all stations and during non-peak boating months.

In many areas of the nation, residents of local neighborhoods have banned together to form volunteer anti-crime "watch" patrols. Such patrols are usually coordinated with local police jurisdictions. Usually two members of the neighborhood patrol will be assigned a shift to circulate in the neighborhhod at night in order to spot suspicious persons, vehicles, or criminal activity. Such things are radioed (usually by CB radio) to the group's central monitoring station so that the police can be called to investigate. There is no single specific CB channel designated or normally used for this purpose, each neighborhood anti-crime watch selects its own primary channel.

Routine anti-crime neighborhood watch activity communications traffic would not seem to meet the FCC requirements for the use of CB Channel 9, although reports of an actual crime in progress, or of a accident or sick or injured person would be authorized on that frequency.

There is nothing to prevent any individual from establishing a Channel 9 monitoring station that will operate on a regularly scheduled or even irregular basis. Naturally, a station having good signal coverage and located near a major highway is preferred. If it is to be staffed by one person alone then perhaps the hours of operation could be limited to the hours of heaviest road use, such as morning/evening "drive time." The base station operator should be able to respond to mobile unit requests for aid that require police, fire, ambulance, or tow truck participation. Therefore, a listing of the telephone numbers for these agencies and services should be convenient to the operator. Since Channel 9 is also authorized for traveler assistance (including providing road directions or information on food, shelter, restroom, or recreational facilities), it would be useful to have available detailed street maps, as well as information on area service stations, restaurants, motels, parks, etc. Of course, it also helps to have a telephone located at the operating position just in case the agency or facility you have to call needs more information from the motorist.

Motorists calling REACT or other emergency monitors should be able to provide all relevant information, including the exact location of the vehicle (highway number and mile marker or distance to nearest exit as well as direction of travel, or the street name and closest intersection or landmark), the specific problem (out-of-gas, flat tire, accident, information about driver or passenger's illness, undetermined mechanical problem, etc.), a description of the vehicle and its license plate number, if any persons are injured and the nature and extent of the injury, how many vehicles were involved in an accident, and if medical attention is required. This is the information police and other agencies will require in order to act promptly.

6

CB QSL's— They're Great!

One of the nicest aspects of CB radio is what CB'ers know as "wallpaper" and the rest of the radio world knows as QSL cards. For the totally uninitiated, here's a quickie/capsulized description. QSL's are personalized station cards made up for use by radio operators (especially CB'ers and hams, although many broadcasting stations have QSL's too).

Ham operators, who used QSL's for decades prior to the invention of CB radio, normally exchange them by mail with DX stations they work. This establishes a confirmation that the contact took place and permits the ham operator to qualify for numerous operator awards that can be earned for contacting stations in a certain number of countries or worldwide radio zones, or states or counties, or with different prefixes, etc.

CB'ers came along and gave an entirely new twist to the use of their QSL cards. While they are used for confirming or commemorating DX contacts, more often than not they are swapped and exchanged in the spirit of friendship between operators, or the sheer enjoyment of building up a collection. Most of the QSL's exchanged by CB'ers are with stations they've never actually contacted over the air-- although sometimes the cards commemorate personal meetings (known as Eyeball QSO's) at CB get togethers, coffee breaks, club meetings, and other social gatheings often held at diners, picnic grounds, union halls, school auditoriums, etc. These gatherings are where new and used equipment is swapped, along with friendship, tall tales, and QSL cards.

Some QSL cards are colorful, others are drab; some are clever and funny, while others seek to mainly present basic no-frills data on the station and

operator who designed the card. The souvenir cards usually contain information such as the operator's name (sometimes including the names of all CB'ing members of the family), mailing address, the CB'ers handle or sideband identification number, possibly a CB callsign. Often the operator's favorite monitoring or standby frequency or channel is indicated. There's usually a photo or cartoon relating to the CB'ers handle or occupation, or perhaps the station itself is shown.

While any neighborhood printer can run off QSL cards made up from your own design, most of the QSL's circulating appear to have been produced by companies that specialize in QSL printing. The classified ads in ham magazines such as CQ and QST show the names and addresses of dozens of QSL printers. The usual practice is to send for samples and a price list-- some samples are free, others ask anywhere from 25¢ to $1. At the very least, I suggest enclosing a large stamped, self-addressed return envelope to any printer who is offering free samples.

Prices vary widely from printer-to-printer and will depend upon factors such as the quality of card stock you want, the number of cards ordered, the number of colors used, and sometimes the complexity of the design. Some printers offer only cards that may be printed from a couple of basic design formats, others invite you to design your own-- or they will custom design a card based upon your description of what you want. Usually the "stock design" cards are relatively inexpensive-- like 100 cards for $8 to $10. Some of the designs that come through are really spectacular, displaying metallic inks and rainbow colors-- of course, the more colorful and dazzling, the higher the cost of the card.

Now that you've been brought up to date, let's get into the nitty-gritty of making the most of these dadblasted souvenir cards. Just some random ramblings I have on the topic based upon my own experiences. Hopefully you can glean a couple of useful morsels out of these babblings since they're intended to help you devise a QSL that has the best potentials for bringing you one in return.

Your Own QSL

Let's face it, some QSL's look so nifty that they scream out for an an immediate response from all who receive them. Obviously, custom designed multi-colored or photographic cards (while perhaps costing more than other types) are going to have a strong appeal. This isn't to say that unless you're able and willing to shell out big bucks your QSL will be a dud without any real "pulling power." You'd be surprised at the really great looking cards that can be devised using only black ink on plain white card stock, or possibly with only one additional color.

Of course, there's no denying that some QSL's are your basic losers, regardless of the number of colors imprinted thereon. Most operators I've spoken to say the cards with the lowest appeal are decked out with crudely or childishly drawn illustrations and lettering, have a smutty or blatant "X Rated" theme, or are done on file cards or paper run through a duplicating machine. Other poorly received cards are those with pre-printed stock designs that are intended for use where the CB'er writes in his/her personal information by hand. And, cards done completely in pencil, pen, colored markers, crayons, or rubber stamps are almost the bottom of the barrel despite the fact that their senders probably consider them to be original works of art. The absolute nadir of the QSL scene is the "card" poorly duplicated on cheap paper that advises the recipient that the sender's "real" QSL will be sent only upon receipt of the recipient's QSL. Such tricks are beneath contempt and usually result in the chintzy stand-in paper "QSL" being given a one-way airmail stamp directly into File 13.

Exchange of QSL's

QSL cards can be sent like regular postal cards and at the rates that apply to that class of mail. For a few cents more you can enclose the card in an envelope and send it by First Class Mail. While you may be able to have a post card arrive in reasonably good condition when it need only to

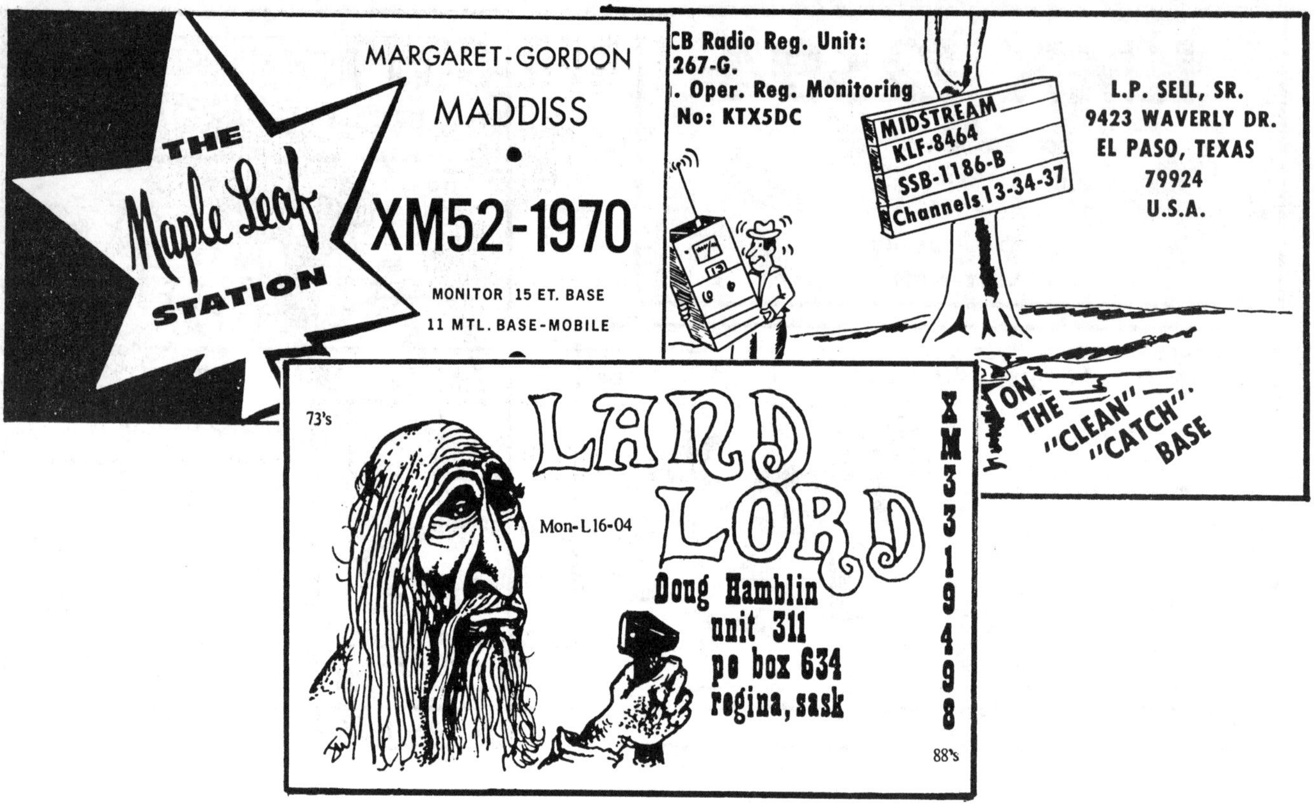

go to another North American operator, for anywhere else in the world I'd suggest going the envelope route so it won't get there looking like Godzilla did a tap dance on it while wearing football cleats.

With postal rates being on the high side, you'll probably find that operators in most nations having the fewest CB operators are somewhat underwhelmed by requests for QSL exchanges from the U.S.A. and Canada. Chances are they've got their houses insulated with zillions of North American QSL's. If it's a country or station you really want, then march down to to a post office and buy an International Reply Coupon (IRC). This coupon should be enclosed with your card so the other operator can use it to buy the stamp needed to send his QSL back to you. By the IRC's aren't of any use to operators in most Eastern European nations. While an IRC won't guarantee a return, but it's a darn good incentive if the other operator is a decent sort. A surprisingly large number of American operators don't seem to realize that American stamps can't be sent to operators in other nations for return postage-- U.S. stamps are good only for letters mailed <u>in</u> the U.S.A, not to letters going <u>to</u> the U.S.A. from Canada, France, Japan, or any other nation.

You might wish to include with your QSL a short personal note and/or a photo of yourself. This is a nice "extra" that impresses many operators. Whatever you do, <u>don't</u> enclose any type of cash "to pay for postage" with your QSL. It has a bad connotation-- like a bribe, a purchase, or a payoff-- and many overseas operators seem to resent and are offended by the practice. I know a Third World operator who tells me that he regularly receives QSL requests from the U.S. that are accompanied by anywhere from $1 to $5 in cash. He returns the cash and the received QSL (without sending his own) accompanied by a polite note explaining that his QSL is sent in the spirit of friendship and that it's "not for sale for American dollars." I'll admit it's odd, but in some areas of the world such things are sensitive matters-- although anybody who wants to enclose cash in an envelope sent to me will find that it will always be graciously accepted.

Of course one nice thing about sending a QSL in an envelope is that

the postage it takes to mail the letter allows you to enclose more than just a single QSL. You might want to enclose several of your QSL's so the other operator can pass them around to his/her friends; possibly this could bring you in return some bonus cards you didn't expect!

Many CB'ers send out their overseas QSL's and letters using the most beautiful and colorful commemorative postage stamps they can locate. This policy is intented to make an impression and rack up points with the other operator in the hopes that he/she is a stamp collector or has one in the family. It's almost a great idea. The problem is that many such festive looking letters never reach their destinations. The reason for this is that the postal systems in many overseas nations aren't quite up to the security standards of those in the U.S. and Canada.

Postal workers in some nations (especially in Latin America) seem to remove such letters from the sorting line and sell them to philatelic dealers. For this reason, it's best to send all mail oveaseas using very plain looking common postage stamps, or (if possible) run them through a postage meter so that they don't have any stamps at all. Radio Vatican, the broadcasting station in Vatican City sends out all of its mail via postage meter just for this purpose. If you want to get your letter there as quickly as possible, spring for the cost of air mail. Sea mail to distant lands is less expensive, but ever so s-l-o-w.

Some operators send what they call a Buddy Pack; that means instead of sending several of their own QSL's, they send one of theirs plus several from other CB'ers. Operators who do this supply other local operators with a stack of their cards to be sent out in the other guy's Buddy Packs. Some QSL's even say "Buddy Packs accepted." Other cards say "QSL 1-4-1," and that means the operator will send you back as many of his own or QSL's from other stations as you send him. And a QSL that says "QSL 100%" (or even, as I've seen, "QSL 99-44/100%) means that he's looking for swaps, and he guarantees to respond to any cards he receives.

Over the years there have come and gone dozens of so-called QSL Swap

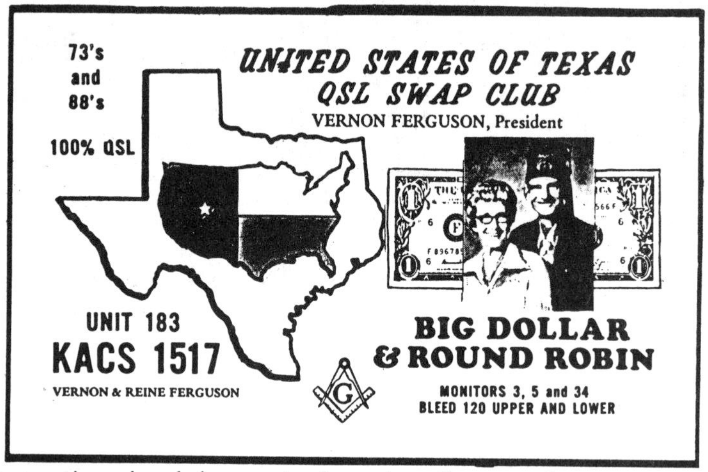

Clubs-- yes, actually national and international clubs catering to those who exchange and collect CB QSL cards. I haven't heard anything about such groups in the past few years, but undoubtedly some are active. When there was a glut of such groups around 1980 or so, swapping fanatics seemed to belong to as many groups as they cound locate. Some groups were actually of genuine benefit in generating QSL's to swap, but others drew very low ratings from swappers-- complaints included sending members nothing at all or else poor quality junk in response to membership fees. The clubs generally promised to send a large Buddy Pack to each new member, plus list that member's name and address in a newsletter going to all other members of the club. Membership fees used to be from $2 to $10 and the new member was asked to supply about 25 (or more) of his own QSL's.

Specializing

As old timers will recall, Chick Sales said that you had to be a specialist (ask your grandpa about it, kid). Sure, you can accumulate as many random cards as you can, but swapping fanatics realize that it adds a lot of challenge to work towards building a specialized collection. Some specialty collections I've seen include QSL's showing state maps, QSL's from members of particular clubs or networks, from particular states or counties, or maybe photo cards, or QSL's showing ships, aircraft, trains, vehicles, etc.

One of the most difficult specialties I've seen anybody try is to see how many consecutive FCC callsigns and SSB club identification numbers could be assembled in a continuous run. Gets rather hairy after maybe five or six QSL's. I have seen a run of eleven consecutive numbers, but it took years to assemble. Old (pre-'70) QSL's are fun, too-- although such cards are really tough to locate anymore and sell for top dollar if you can locate anybody who is willing to part with them.

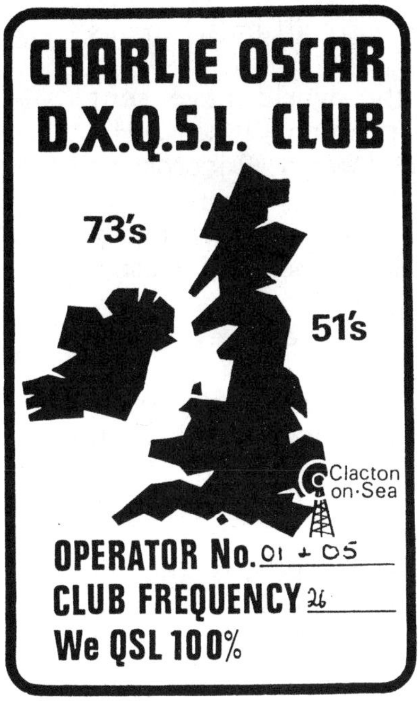

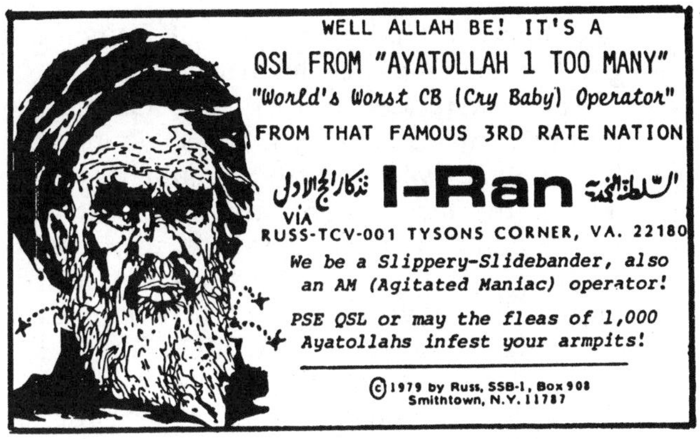

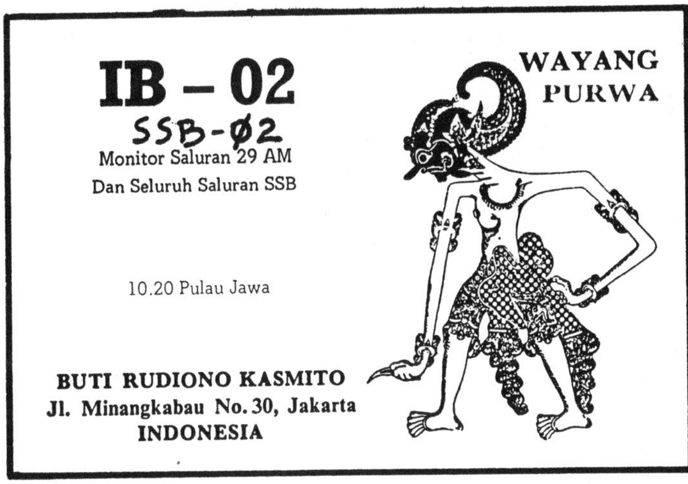

Display/Storage

You can toss all of your cards unto a cardboard carton, but why not do something creative with them? Arrange the QSL's by state, nation, call sign, handle, SSB numbers and you can put them in shoe boxes, or hang them on your walls, or place them in albums or scrapbooks. I don't suggest attaching them to walls by means of thumb tacks because it makes the cards look ratty and doesn't do much for the surface of the wallboard or paneling either. Nor do I suggest attaching the QSL's to anything by means of any type of tape or stamp hinges or any other gluey or sticky methods-- really messes them up.

Best way to put QSL's on your wall without ruining the cards is by hanging large sheets of oaktag on the wall, then attaching the QSL's to the oaktag by means of those little triangular "corners" intended for mounting photos in albums. Use two on each QSL, placing them at opposite corners. Don't ever display QSL's or store them near extremes of high or low heat or humidity. Poor locations include basements, cellars, attics, garages, or near heating devices.

A well prepared and displayed QSL collection always makes a hit at CB gatherings-- but when displaying them there, cover them totally with a large sheet of plastic to protect them from spills, burns, handling, and fluttering to the floor. Keep your eye on them at all times!

International Friendship via QSL's

CB'ers have always been generous souls. They're already to organize fund raising drives when one of their own is in need. Countless persons in need have been the recipients of the good works and desire to be of service that has been a hallmark of CB since its earliest days. But sometimes, things get out of hand-- at least a little.

For instance, since 1982 CB'ers from every part of the world have sent millions of QSL cards to a young Scottish boy with a terminal illness. His final wish was to find a place in the Guinness Book of World Records as having the largest QSL collection.

It can't happen. First of all, the Guinness book doesn't record a category relating to QSL collections. Also, the dying Scottish boy, Buddy, doesn't appear to have ever actually existed. The entire thing seems to be a highly complex hoax that touched hundreds of thousands of CB'ers and even the President of the United States!

About 600,000 cards per year arrive for Buddy at his supposed hometown, Paisley. As of 1987, 3-million of them have shown up. Even though the British Post office has devoted considerable effort to telling everybody who would listen that it's all a hoax, still little Buddy receives more mail that the British Prime Minister and Royal Family combined.

Seems that the hoax began when a Paisley CB'er heard via CB that a child, Buddy, was desperately ill and was collecting QSL cards for a world record. The CB'er began spreading the word, telling other operators to send their cards to his own postbox so that, when Buddy's address was obtained, the cards might be forwarded to the child. After the first three months there were more than 185 sacks of QSL cards on hand and Buddy couldn't be located despite diligent checking with schools, children's homes, and hospitals all over Scotland.

From the return addresses on the QSL's, it seems that the cards originally began arriving from the British Isles, then North America. Within a year Buddy had even received a cheer-up letter from President Ronald Reagan.

After leaving North America, the plea-for-QSL's rumor advanced towards Australia, then to Western Europe. More than one major airline pilot requested international passengers to send Buddy get-well cards, or QSL's if they had them. Eventually, when the traveling tale returned, via Scandanavia, to the United Kingdom, Parliament began preparing a resolution that suggested citizens to send mail to the child.

The British Post Office was able to convince the lawmakers that it was all a hoax that had run wild, anthough postal workers long ago stopped seeing any humor in the matter. QSL's, get-well cards, letters, presents, cash donations, and all sorts of other items for Buddy continue to arrive by the thousands each week and showed no signs of slowing down-- even after five years!

Although most of the original mail eventually went through shredding machines, of late the mail is distributed to childrens hospitals where the stamps are removed to be sold to philatelists. But, basically, the British Post Office really would prefer if CB'ers would stop perpetuating this story so that the QSL's and other mail for the imaginary Buddy would finally grind to a halt.

There is every indication that good-natured CB operators will continue to do whatever they can to help this terminally sick child to achieve his last wish-- even if it takes decades!

7

Single Sidebanding

What are your communications needs? Do they include reaching out from your base or mobile station in order to talk to stations as far away as possible, given CB's power limitations? What if you could, legally (of course) operate on all CB frequencies with an effective power output several times that used by folks who run the standard 4-watt transmitters, thereby granting yourself a communications range that can make your give your mobile unit more range than some base stations, and permit your base station to have groundwave coverage as much as fifty or sixty miles? Nice, eh?

You can do it, all you have to do is climb aboard the single sideband (SSB) bandwagon and join the crowd! Some folks refer to this crowd as "Silly Sidebanders," others call them "Sensible Sidebanders." Many people (including

Are Sideband operators snobs? It's open to some debate. Certainly they have cultivated a special standard of operating practices that looks down on CB "handles," 10-Codes, CB jargon, and just about everything else dear to the hearts of most CB'ers who operate in AM mode. Some Sidebanders don't mind one bit being called snobs.

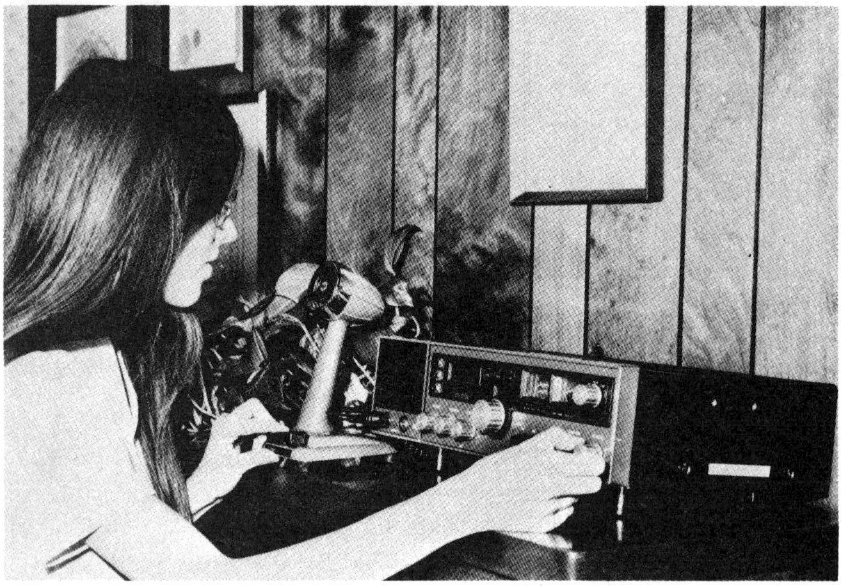

Sidebanders are quite happy in their own little corner of the CB band. They ask nothing more than to be left alone to do their thing on frequencies that are an oasis of peace, quiet, and conversation.

some I've spoken to at the FCC) think that they may well hold the promise for the future of all voice communications on CB. Many active CB'ers barely even know they exist!

You might say, "What's the big deal? Single sideband (SSB) has been around for decades-- it's just one more variation on getting a voice signal from here to there." From a strictly technical standpoint, this might be correct, but on 27 MHz, SSB is a whole different ballgame.

Some who are an active part of the sideband movement hesitate (or even react in horror) at calling themselves "CB'ers," so different is their world from the world of the majority of regular (AM mode) CB operations. Many of these people refer to themselves as "11 Meter Sideband Operators" or something of a similar nature. They are most definitely not in the same boat as the vast majority of CB operators, and they aren't at all bashful about pointing this out to all who will listen.

And rightly so, for 27 MHz sideband is far more than a technically different method of sending a message. It's more of an emotion, a state of mind, a hope, and maybe even a dream! And those who are part of this 11 meter band sidebanding fraternity are often rather militant in their defense of something they feel is (or might be) ripped off or plundered by what they consider hordes of screaming, shouting, Smokey-reporting, 10-4'ing, CB'ers. Sidebanders, in fact, sometimes tend to feel that they are the CB elite.

Are 11 meter sideband operators snobs? Certainly it can't be denied that they are often clannish, some even say straight-out that they're snobs!

I say not so! I see these people as those who have a pretty good thing going and want to see it maintained at a reasonable, if not high, level of usefulness and rationality.

The Typical Sidebander

The typical Sidebander is a person who started out with an AM CB rig and then found, for one reason or another, the AM mode of operation was less than satisfying. Perhaps disenchantment set in of too many stations jamming the channels, too many instances of getting walked on, too many breakers and heterodyne whistles, too many signal chuckers with their dead carriers and attempts at playing music on a busy channel to relieve some operator's hostilities towards the world-- just too much a wall of humanity and signals clattering away on the channels-- folks asking for radio checks, many time checks, and using CB handles. True, some folks-- maybe most of those on the CB channels-- love this carnival-like atmosphere, and those people

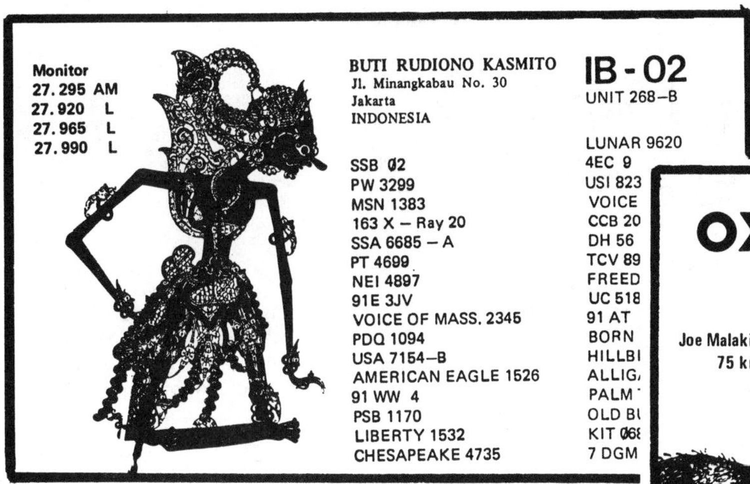

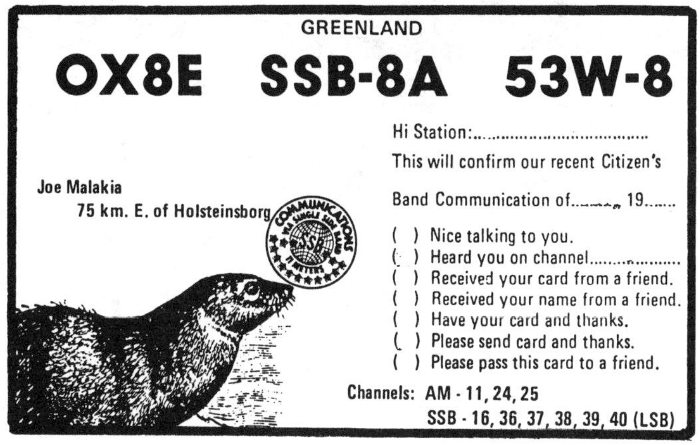

seldom gravitate to the world of Sidebanding. In that world, fewer stations and less activity make communications more pleasant, and where those who are there go out of their way to maintain an orderly and somewhat structured communications system.

Simply having a sideband operating capability on CB rig doesn't automatically and necessarily make an operator a Sidebander, not by a longshot. Or, as I once heard a Sidebander coldly tell an operator who blundered by asking for a "10-36" (the correct time), "Just because you're head is shaped like a banana doesn't mean you're one of the bunch!"

Some Sidebanders don't operate on AM at all. It seems to me though that although the majority of sideband operators do use both AM and SSB modes, the AM being used primarily for monitoring Channel 19 for road conditions while mobiling. The operator who mostly uses AM, but who clicks over to sideband on rare occasions "just to see what it is," who knows or cares nothing of what Sidebanding is all about, is not considered a member of this close-knit fraternity-- not that serious new members aren't always welcomed with open arms. And woe unto the poor AM operator who attempts to transmit in AM on any of the channels staked out by Sidebanders-- they run the poor station off like a screaming eagle protecting its nest!

So the typical Sidebander is probably a person who uses that mode most or all of the time, not only because of the extended communications range, but also because it offers something totally different than what AM activity

One of the factors that has kept Sidebanding on the CB band less crowded than AM is the relatively higher cost of the equipment required.

It takes more than unpacking a Sideband rig from the box and plugging it in to make a person a Sidebander. It means learning the different procedures and disciplines the operators demand.

has to offer. Thus the 11 meter Sidebander is usually a rugged individualist whose rewards are many and that suits him/her just fine.

Why are there fewer Sidebanders? One reason is that most CB operators are quite content with, or are at least willing to tolerate AM operating conditions and don't wish to become self-exiles into what they fear may be some sort of exotic never-never land. Of course, many AM operators know so little about Sidebanding that they aren't even ready to consider the possibilities. Also, CB equipment that can operate in SSB mode (even though it can also operate in AM mode, too) costs more than AM-only CB transceivers. In any event, even though the ranks of Sidebanders are growing, there are fewer Sidebanders than AM'ers.

Joining The Fraternity

It's difficult, if not altogether impossible, to purchase a sideband rig and expect to unpack it and immediately be accepted into the fold, as happens with AM operations. Sidebanders have their own lingo, their own traditions, their own unique levels of tolerance for what they will and will not allow on their channels. Those who would commence operating on a sideband channel, even after many years of seasoned operating on the AM-side, will soon come to learn that a great many of their most cherished operating procedures, buzzwords, codes, and expressions are totally unacceptable in this new world they seek to enter. Thus, Sidebanding may mean a reeducation that involves un-learning what may have been acquired over years of experience.

A new operator who is sincerely trying to integrate into Sidebanding is always helped and even given suggestions by those on the channels. However one who persists on using AM practices on a sidebanding channel soon learns that it was a mistake to do so, for a well-deserved razzing and needling with little mercy is dished out. The erring operator either learns and straightens out or simply writes off sideband operators as a bunch of elitist, uncooperative cranks and forgets about the idea of communicating with them.

Rudeness, for instance, is a definite no-no. Any attempt at identifying one's station by means of a CB handle on a channel normally used by Sidebanders invariably unleashes a hornets' nest of outrage that stops only inches short of shouts demanding that the head of the infidel be passed around on a silver tray!

So, yes, there are definite rules of the road. Still, learning them and become a part of Sidebanding opens up a whole new world of enjoyment and friendship-- and learning the ropes of Sidebanding is not a particularly awesome task. Those who can't or won't learn those techniques, and who return to AM operation, only serve to keep Sidebanding the way the operators wanted it in the first place.

Sidebanders, after all is said in done, are really just interested in communicating. Taking advantage of the fact that there fewer stations crowded onto the channels where Sidebanders congregate, they are able to get involved in roundtable discussions on a wide variety of interesting subjects-- often (but not always) on the art of communications and the latest antennas and transceivers for 11 meter use. The operators are, in every sense of the term, friends. Therefore, their conversations cover the kinds of things that friends might normally discuss. There is a general aura of courtesy, consideration, and gentle humor in evidence on the frequencies.

Moreover, the extreme long distance operating range of the equipment permits the SSB'er to include within his/her circle of friends operators in a far wider coverage area than would be dreamed of on AM-side. In the event of a roadside emergency, when out of range of a Channel 9 (AM mode) monitoring station, it's entirely possible that the only stations that could be reached to send help are Sidebanders operating on frequencies where these stations normally operate.

In general, it's a laid back and casual opportunity to pleasantly communicate with a great group of people, all of whom share your enthusiasm for a mature approach to communications.

Getting Technical About Sideband

To many people, SSB is a somewhat vague and exotic technical advancement of the Space Age. In actual fact, it's more than fifty years old, and it's not especially mysterious. An explanation of its basic is well worth the time of anyone seriously thinking about becoming involved.

SSB is a superior means of carrying intelligence, superior to either FM or regular AM, but it doesn't really make the other systems outmoded. AM and FM certainly have advantages over SSB in certain applications, but those advantages must be evaluated in the light of the use to which these systems are to be put. It is ironic that some of the advantages of AM and FM are not only unimportant to CB-type land mobile communications, they

A brief explanation of what makes single sideband different than AM mode is rather painless, but goes a long way towards pointing out its many advantages over AM mode.

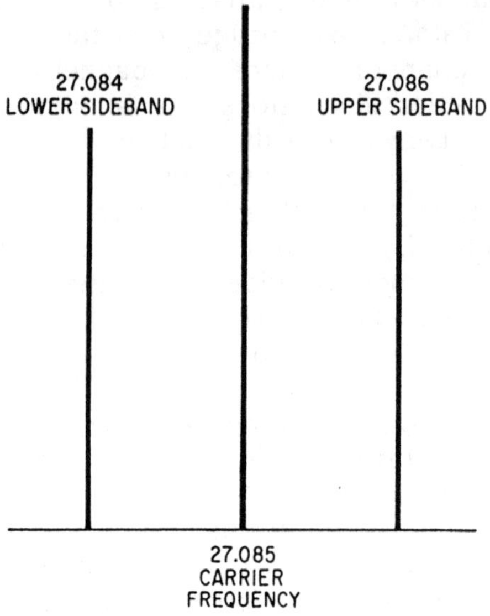

An AM signal, consisting of a carrier and two sidebands.

are actually disadvantages-- even though, in recent years, there has been some use of FM-mode by stations (unauthorized in the U.S.A.).

A simplified explanation of SSB involves the basic principles of both AM and SSB. When you add voice power to a carrier generated by a standard AM transmitter, two "audio sidebands" develop. These sidebands are the sum and difference frequencies, plus and minus the frequency of the carrier (lets say 27.085 MHz), with the audio frequencies added or subtracted from the carrier frequency.

In other words, Channel 11 is 27.085 MHz. By adding a simple 1 kHz audio tone from a tone generator or other audio source, we immediately have three frequencies emerging from our transmitter. One is the 27.085 carrier frequency. Another is 27.086 (1 kHz-- the audio frequency-- above the carrier) and another is 27.084 (1 kHz-- the audio frequency-- below the carrier).

Why do we have these frequencies? Because, simply, the final amplifier stage in the transmitter does two things. First of all, it develops about a 5-watt input signal. Second, it acts as a meeting ground for the 5-watt input frequency-controlled signal and the audio which develops in the modulator

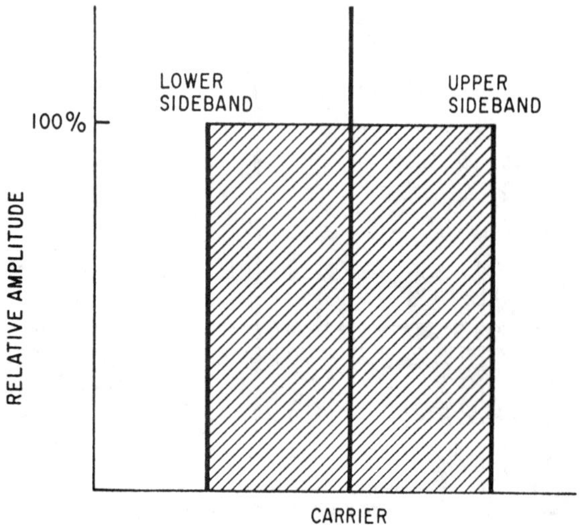

On AM, the full "talk power" is spread out over the carrier and both sidebands.

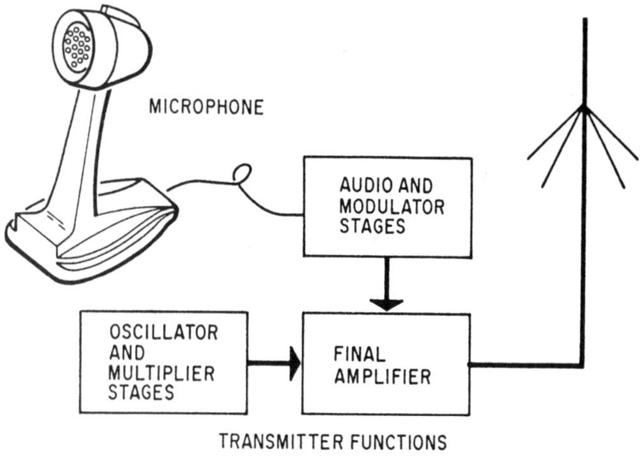

Block design of an AM transmitter.

stage when you speak into the microphone. The audio is added or applied to the transmitter carrier in the final stage of the transmitter. The transmitter final amplifier is already on our example frequency of 27.085, or Channel 11. The voice power from you, however, is in the human speech range, or 300 to 3,000 Hz (0.0003 to 0.003 MHz). The 27.085 MHz signal and the audio signal mix together in the final stage. The audio (voice) frequencies add to and subtract from the carrier frequency at a rate dependent upon the changes in the voice speaking into the microphone.

In SSB operation, we take advantage of the fact that it is not the carrier that contains the intelligence (i.e., the voice) but rather the sidebands that are actually displaced from the carrier frequency by an amount equal to the frequencies of the voice.

The carrier, then, is nothing but that-- a carrier. And yet it doesn't carry anything! Once we have established the frequency of our transmitter with the carrier, and mixed our voice intelligence with that frequency, thereby developing our sum and difference frequencies, the carrier itself is of no more use to us until the signal arrives at the receiver. The carrier, then, is not needed to transmit the voice from here to there. It is required only to develop the proper transmitting frequency for the signal.

So we can ignore the carrier. We'll simply filter it out of our output, leaving us with two sidebands, the upper sideband (USB) and the lower side-

Audio plus carrier relationship.

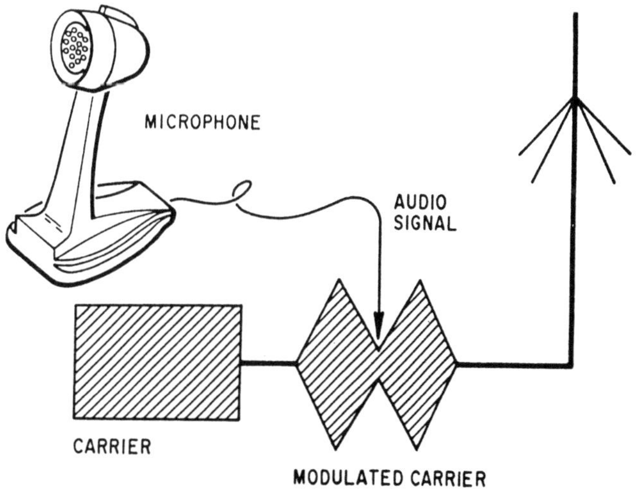

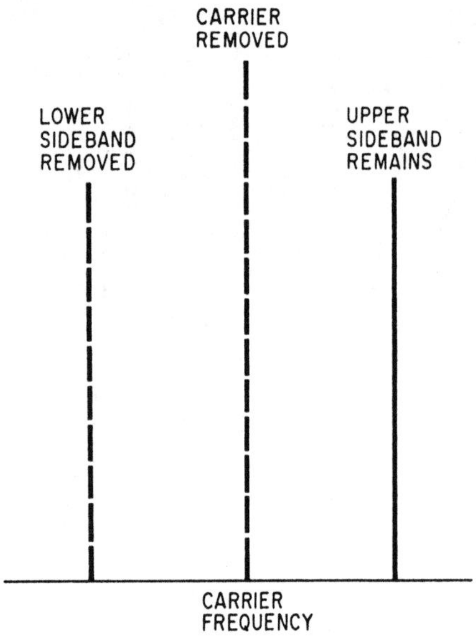

On SSB the carrier and one sideband are eliminated.

band (LSB). Filtering out the carrier is done by one of several electronic methods which aren't important to worry about in this basic discussion.

Of the two sidebands remaining, each contains the same intelligence. Should we then be able to ditch one of them? Of course (I'll explain why shortly.) Once again, a bit of electronic filtering is employed which wipes outout one of the sidebands-- whether it is the USB or LSB is irrelevant so long as the communicating stations have agreed upon which sideband will remain. There is no technical advantage or reason to using one over the other since they are both identical-- although 11 meter band operations generally take place in LSB mode.

What we have gained by filtering out first the carrier and then one of the two sidebands are 9 decibels (dB). Or, to put it another way, we have increased our effective transmitter power output eight times. This is assuming that the transmitter is operating at the 5 watt input level (which yields about 4 watts output in AM mode). The sideband power rating for the CB transmitter is rated at 12 watts (PEP)-- and it's still within the FCC's power limitations for CB stations!

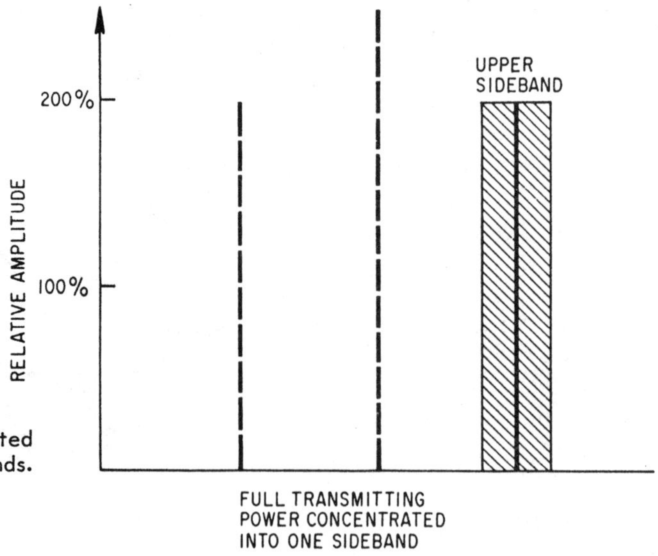

With SSB, the full "talk power" is concentrated into one of the sidebands.

With the carrier extracted, all that's left is the message itself.

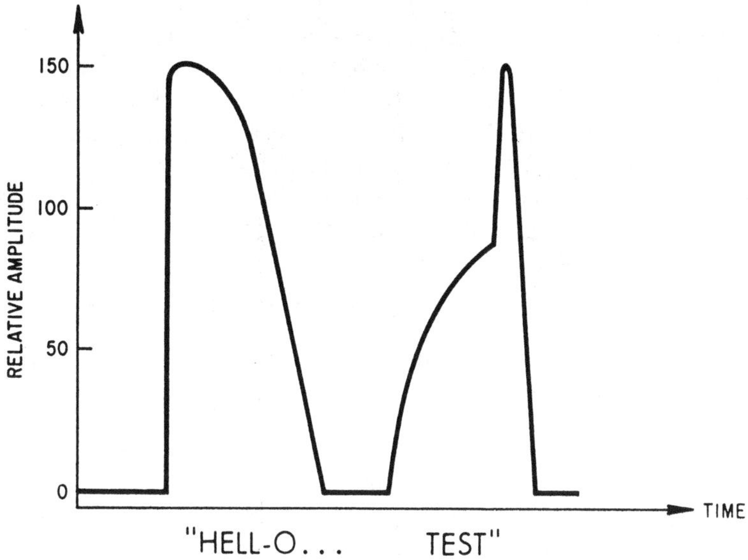

Why have we gained this power advantage by using SSB instead of AM? And how can we still run 5 watts input and yet increase our effective input to the point where an AM transmitter would have to run 40 watts input (32 watts output) to produce a signal of equal potency?

Here the theory may get a bit involved for the newcomer. Let's simplify it this way. The 5 watts previously in use with AM was spread out over three areas-- the lower sideband, the carrier, and the upper sideband. By knocking out the carrier and lower sideband, and still maintaining our input power to 5 watts, we have all of the power to dedicate to being concentrated in the remaining sideband. In other words, we aren't diluting and wasting power by squandering it on an extra mirror image sideband that is needlessly duplicating what we are saying over the first sideband.

Power Measurement

On SSB we measure power a bit differently from the way it's done on AM. With an AM carrier we simply turn on the transmitter, pipe the signal into a wattmeter, and watch the indicator needle zip over to show the power output.

On sideband the only time there is any power output is when we are talking or feeding some type of audio into the transmitter. This means that every time we say "hello test" the needle on our wattmeter bounces around like mad. On "hello" the output suddenly jumps from nothing (remember, no voice-- no output) to some value on the meter scale. Between the words "hello" and "test" there is a slight pause-- so the transmitter output briefly stops-- actually shuts down completely until the word "test" is spoken. In these two words the meter also hits various peaks when one's voice inflections are added to a particular letter or sylable, if one word is more accented than another, or the sound is a whisper or a shout. The illustration shows how a typical power output meter reacts as we say the phrase "hello test."

As you might imagine, in a twenty-word sentence, the output is going to fluctuate tens of times. So what do we measure-- the peak power hit on the meter when the accent is heaviest on the longest word or the power output between words (zero output)? The answer is neither. We measure the average of the two.

This means that for any given sentence or transmission we will have peak power considerably above our average power. Mathematically, it figures out to be twice the average power.

Remember that 5 watts of input power? That was average power. Our peak power is now 10 watts. FCC regulations, taking into account that occasional modulation accents may exceed this, state that the peak effective power (PEP) can run up as high as 12 watts.

All this by way of explanation for those who were wondering where the 9 dB came from. The same 9 dB advantage that we gained when we filtered out the carrier and one of the sidebands in our AM transmitter ended up in the single remaining sideband. It's the electronic filters, doodads, and gizmos needed to do all of this magic that makes an SSB transceiver cost more than an AM-only CB rig.

Does 9 dB Mean Anything?

It just crossed my mind that unless you are familiar with signal measurement techniques and ratings, the revelation that there is a 9 dB gain may not mean much to you.

Engineers know, through agreed-upon principles, that when the signal doubles in strength, the increase in level can be detected by the human ear. This, however, is the smallest change that the average human ear can detect. In other words, if you double the amount of signal you are putting in at a distant receiver, the listener will detect the change in level as an audible improvement in your signal strength. This doubling will require that you either double your transmitter power or add on an appropriate amount of antenna gain, or perhaps add a combination of things.

I had said that 9 dB of signal increase was the same as using eight times

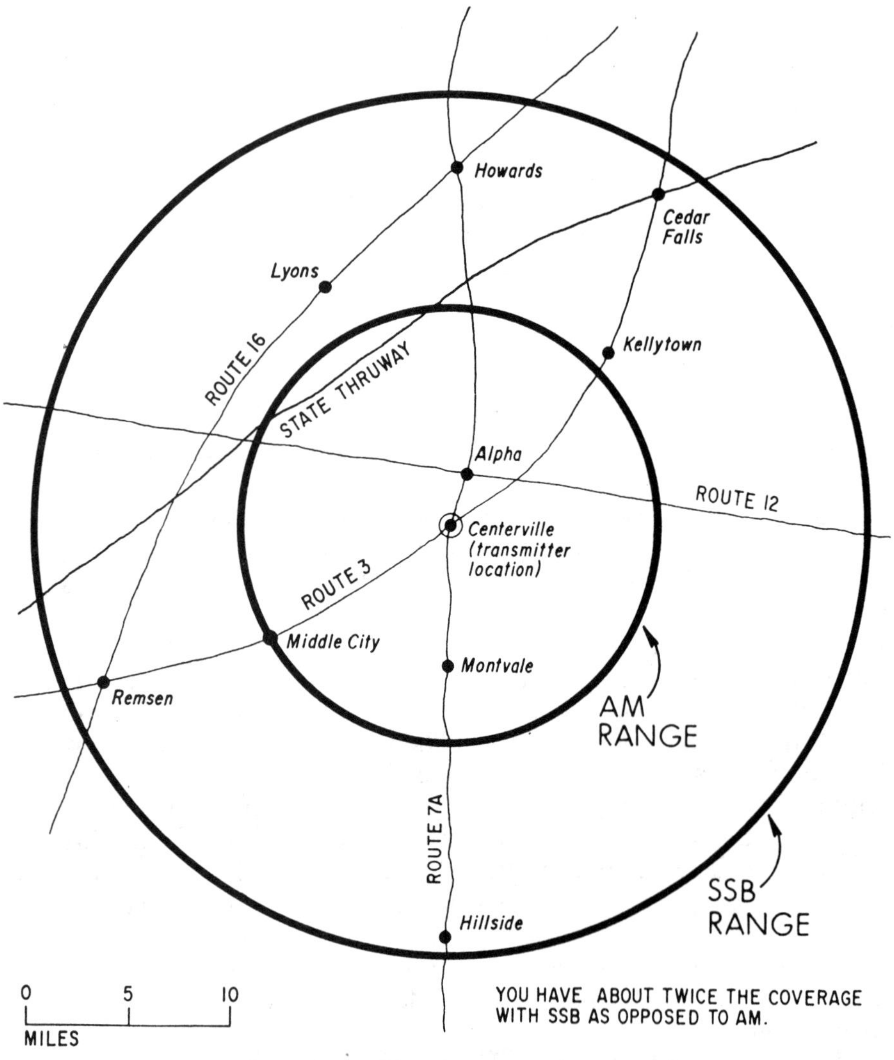

Increased range of SSB.

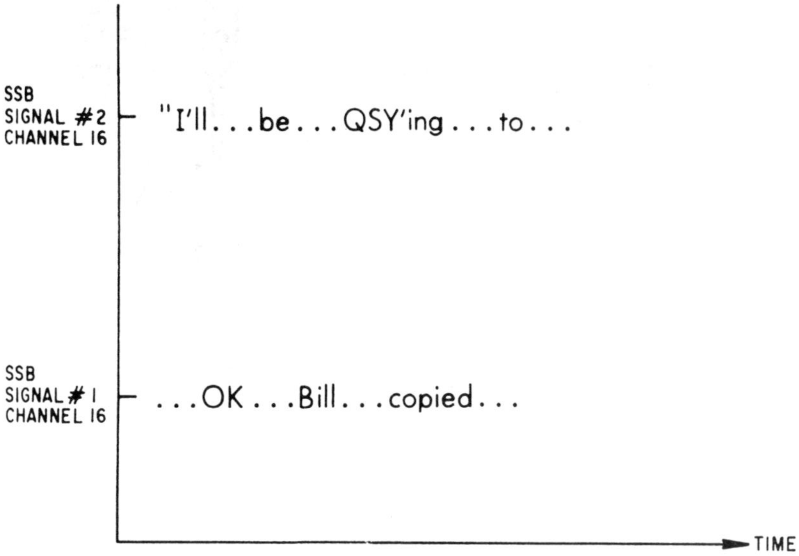

Two SSB stations could occupy the same channel simultaneously, although it isn't very practical.

as much power. Eight times the power is like doubling (1+1=2), then doubling that again (2+2=4), and then doubling once more (4+4=8).

Engineers measure a doubling of signal level as "3 dB of signal improvement." This doubling three times (from 1 to 8) is actually the same as 3 X 3 dB -- that is, 9 dB. We have made progressive increases in the minimum detectable amount of audible signal change over our first signal level.

That is 9 dB.

Advantages of SSB

From a practical standpoint, let's see what all of this means-- the actual advantages and disadvantages of SSB in an communications situation.

The same 9 dB we were talking about gets the spotlight as far as advantages go. A 9 dB circuit gain (that is, the extra signal level we can expect using an SSB CB transceiver as opposed to AM at 4 watts carrier output power) can be very important in the fringe of your present coverage area. If you are currently experiencing 10-mile base-to-mobile coverage, where the limiting factor is primarily the fault of a weakening signal level and not ignition noise that drowns the signal at the 10-mile point. That extra 9 dB of effective signal (some call it "talk power") will add from about 5 to 8 miles to your base-to-mobile range-- not quite double, but almost double your present range. The same ratios can be applied to any circuit distance you are now working. Spot your base on a road map of your area. Now measure the distance on a line from your base station to your farthest regular mobile or base contact point. Extend this same line by 50% to 75%, and then draw a new "SSB coverage" radius using the new extended distance.

Using this simple technique, you can quickly see the fringe area coverage value SSB can have. Of course, your actual station coverage probably doesn't show up on a map as a perfect circle.

Other than gain, SSB offers an additional advantage or two. Let's go back to the point about filtering out the carrier. With the carrier gone, there is no constant signal on the air. Only when we are speaking is there any signal present, then only during word syllables. Even between syllables of words there is "dead air time." When you read the following sentence aloud, as fast as you can and still have a listener understand you, you will have had more than 50% of dead air time just between the words and syllables. Try reading, "OK, Bill, copied your last transmission all the way and in the clear."

With the carrier gone, all that's left is the message-- that is, the words.

Think back now to the last time you heard two AM stations on the same channel at once. In addition to the stations, you also heard a shrill squeal or tone (called a heterodyne) at the same time. The tone was caused by the two carrier signals from the AM transmitters trying to occypy the same place in your receiver at the same time-- something that isn't accomplished with ease. Also, the odds are something like 100,000 to one that the two transmitters you heard were actually not on exactly the same frequency. The FCC rules say that transmitters may deviate from the assigned channel ("center slot") by up to 0.005% of the channel frequency. The difference between the two carrier frequencies may have been a matter of only 200, 400, or 800 Hz. But that difference, when the two signals were mixed together in your receiver, came out of your loudspeaker as an audio tone of 200, 400, 800 Hz, or some other random note frequency between 20 Hz and 4 kHz. This is a beat note, so called because the two signals are beating together. Even when you are listening to another SSB station on your receiver, the pollution of the channel by signals from an AM transmitter will cause an annoying heterodyne (as the AM signal mixes with the SSB detection circuitry in your receiver), which is why Sidebanders usually prefer to operate on channels

that they use on an exclusive basis. Despite numerous requests from SSB'ers for the FCC to designate certain CB channels for exclusive sideband use, the agency has refused. So far as the FCC is concerned, all 40 CB channels are open to use by all stations, AM and SSB. Separation of the two modes in any given local area has therefore been left to mutual cooperation between users of the two modes, usually with varying success-- for SSB stations operating on channels where AM stations are trying to communicate create their own unique interference.

On SSB-only frequencies, there are no heterodynes or beat notes since there aren't any carriers to beat against one another. In other words, on SSB it is conceivable (although not actually practical) for two or more conversations to be going on via the same sideband of the same frequency in one local area.

Another thing to consider is that, compared to AM, your SSB receiver is picking up only half the spectrum space. So if you are trying to live with a high ambient noise level, you are now slicing out half of the noise level that you would have had to live with on AM.

Here's a bonus-- most AM rigs draw about 1.5 to 2 amperes when transmitting, and a tad more when you actually speak into the microphone. With

The SSB Network's "Lucky Buck."

an SSB rig there is far less power consumption, less than 0.25 when you are not speaking.

There are a few things that are a burden upon the SSB operator, as compared to the AM-only operator. With the cost of SSB equipment being considerably more than AM-only rigs, that's an obvious factor.

Another factor is that there are some additional operating considerations in SSB that AM operators don't have to worry about. The SSB operator must select (by front-panel switch) the appropriate sideband for the desired channel-- USB or LSB. Also, in order to produce an intelligible signal in the receiver, the SSB operator must adjust a clarifier control on the front panel of the transceiver. This will probably have to be slightly readjusted for each station worked. It's a fine-tuning control, and unless the stations being received are clarified in, they will tend to sound tinny and like voice synthesizers. A variation in transmitting frequency of only a few cycles (Hz) among different SSB stations on the same frequency will be noticeable in an SSB receiver.

We are faced, then, with the additional "problem" that voice fidelity in SSB isn't equal to either FM or AM communications systems. At best, even with maximum clarification adjustment, SSB signals don't reproduce the voice with the same naturalness as other communications systems. But, after all, we are primarily trying to accomplish transferring messages from here to there. We aren't really attempting to achieve the type of voice quality that would be sought by a broadcasting station. Taken on balance, this loss of some voice fidelity is a small enough trade-off to make for SSB's numerous other advantages.

All in all, SSB offers an almost ideal answer to CB communications within the current limitations of technology and the FCC's technical standards.

Understanding & Buying SSB Equipment

If you've ever gone out shopping for CB gear you may well have come to the conclusion that there's something unique about sideband equipment: it costs more than AM-only gear. This is one reason why sideband hasn't grown in use at the same rate as AM. But most sideband operators aren't unhappy about the fact that they are fewer in number-- feeling that quantity isn't as important as quality. And it's the quality of the gear that has kept the prices higher than those of standard AM gear. It's doubtful that you'll ever see any sideband gear selling as low as $65, as is the case with AM sets.

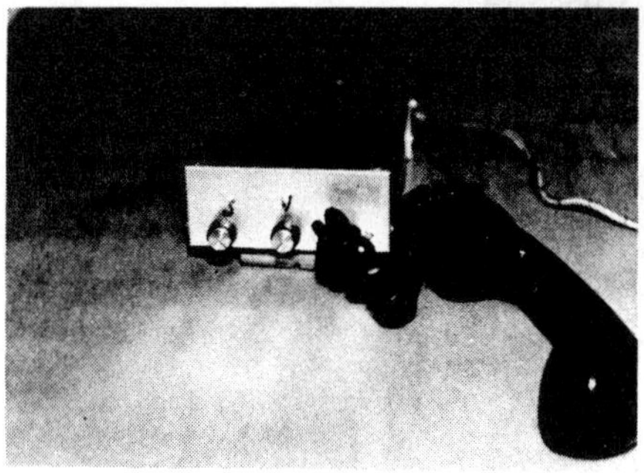

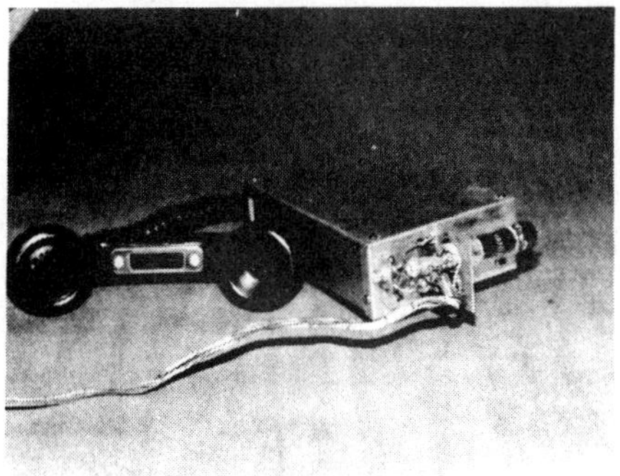

First Known CB SSB Experiment

Probably few Sidebanders are aware of the 1961 CB SSB experiments made by George L. Mason (Amateur W3IZC) of Washington, D.C. So far as anybody can tell, nobody else did it before George. Finding that his AM CB coverage was too limited, he wondered if he could extend his range by using SSB. Of course, in 1961 there weren't any CB SSB rigs on the market, nor did the FCC rules specifically mention that SSB was allowed on CB frequencies.

He asked the FCC and they said it was OK, but they told him to make certain that he stayed within the CB regulations. In those days, CB'ers were permitted to have homebrewed equipment, so Mason designed and built a tube type base station with a transistorized power supply (cost: $500). Frequency stability was achieved by means of a small crystal oven operating at 180° F. A similarly designed mobile rig was also built. The base station antenna was located atop a 2-story building and gave 100% copy at 25 miles, with some coverage at certain locations out to 50 miles. These units operated until 1964 when they were replaced by homebrewed transistorized (except for tube finals) rigs-- those were used until 1970. Nice going, George!

Since, by its very nature, SSB requires more exactingly designed and produced equipment than would be needed for AM-only operation, the current crop of sideband gear is the result of a number of years of evolution, with designers constantly striving to improve the technical aspects of their products. Of course, the fact that sideband gear usually can be switched into AM-mode operation, too, is a big extra for your money.

Actually SSB gear got off to a pretty good start, for the ancestor of modern CB sideband gear was a rig so high in quality that, even though it was produced in limited quantities back in 1964, there are still some in actual operation and sounding as good as any of its modern counterparts.

Since old-time Sidebanders constantly speak of various legendary pieces of equipment, a few historical facts might be interesting to newcomers. The first commercially produced CB sideband rig was the result of a project started in early 1963. Mark Products, a division of Dynascan, hired Gil Kowols to design a CB sideband transcriver. Gil had the prototype version completed in March and (using his ham callsign) he promptly tried the unit on the 10 meter ham band; seemed like a good idea since there weren't any sideband stations yet operating on CB. The ham band shout resulted in a reply from a station in Panama!

One of Kowols' sideband rigs was exhibited at the 1963 May Parts Show in Chicago. As the unit was readied for production, it was awarded U.S. Patent 3,300,720 covering the transceiver and the frequency relationships. Preproduction commenced in July of 1963, and twelve experimental units were produced. These were field tested and demonstrated by the company's sales force.

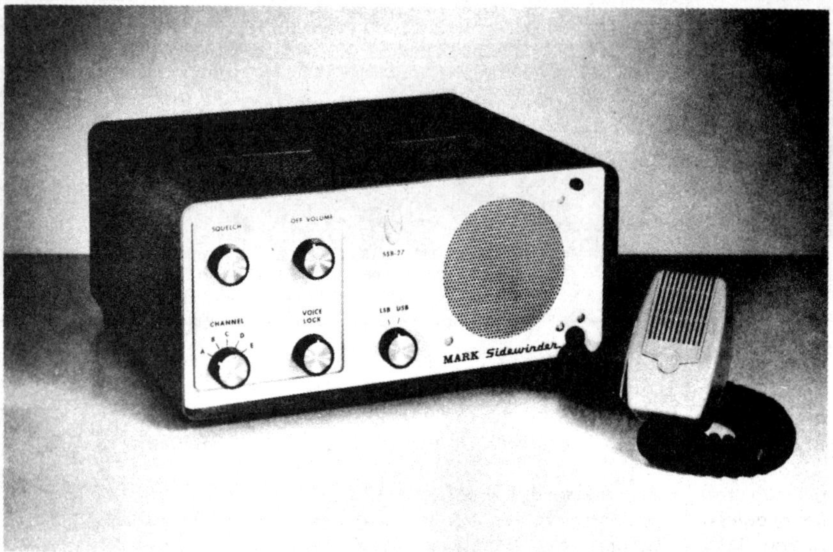

Mark Products designed and marketed the first commercially available CB SSB rig in 1963. It was called the Mark Sidewinder SSB-27.

During these field tests it was discovered that ignition noise and other similar types of noise had a profound effect upon the range. As a result, a noise blanker was designed and was offered as an accessory. The set was dubbed the Mark Sidewinder SSB-27 and placed on the market.

In its final version, the Mark Sidewinder offered selectable USB/LSB operations on five crystal-controlled channels. It had seventeen tubes and many other features, such as a four-section crystal lattice filter, a crystal oven for frequency stability, crystals cut to 0.0015% accuracy, and a solid-state power supply. An optional base station power supply was available.

An unfortunate decision was then made by the manufacturer. Instead of selling the rig via the large network of antenna distributors that Mark had established for their Heliwhip brand CB antenna line, they decided to market the SSB-27 through the outlets that handled their line of TV test equipment. This, it was hoped, would permit the set to be handled by retail giants such as Allied Radio and Newark Radio. The small CB specialty shops would not handle the rig.

By the end of 1965 less than 500 sets had been sold, and there were major corporate changes being made, one of which resulted in Ed Harris, the former General Manager at Mark, leaving to establish his own company. He purchased the remaining SSB-27 inventory and directed his sales efforts at the CB specialty market under the company name of SSBCO. Harris also brought in engineering talent to add AM capabilities to the unit, and thus was born the ASB-11, in a production run of about 500 units. This was the first true AM/SSB transceiver.

The majority of sales were in the Midwest, where many of the operators who had purchased the units started a club called Horizontal 16. These operators had combined sideband operations with flat-side (horizontally polarized) antennas. All members used SSB-27 or ASB-11 rigs.

Wherever these rigs were sold, those who had them soon formed tightly knit clubs and networks. Sideband seemed (and still seems) to produce a

One of the Mark Products executives purchased the remaining unsold Sidewinders and started his own company devoted to selling the units in the hopes of cultivating the single sideband market.

The Johnson Messenger 350, affectionally called the Johnny 350. It was a two channel SSB rig made back in 1967.

feeling of fraternity among its enthusiasts. On the East Coast, owners of these rigs established the well-known SSB Network, which still exists as the oldest, largest, and most influential network active in Sidebanding. Among the founding members of the SSB Network was rock music star Gary U.S. Bonds (SSB-17), who was alternately knocking out gold records and a whale of a sideband signal. I was also one of the original members of the SSB Network. Interesting, almost all of the original members of the SSB Network are still active Sidebanders!

Although it appears that it was the Mark Sidewinder SSB-27 that got the ball rolling, it didn't remain on sale long enough to maintain the momentum. The next step in the evolution of CB sideband came in 1967, when the E.F. Johnson Company introduced the Model 350 Messenger. The Johnny 350, as it was popularly known, was a solid-state 2-channel rig that came factory equipped for operation on Channels 16 and 18. The fact that the Johnny 350 wasn't a commercial success wasn't the fault of the equipment. It was an extremely fine piece of equipment, but there was too little information about sideband available to the average consumer to allow the unit to achieve the popularity it deserved. However, enough did reach the public to stimulate an an irreversible spark of growing interest in CB sideband.

Eventually manufacturers such as Tram, Browning, Stoner, CPI, Cobra and others produced sideband transceivers that live on in both legend and also in actual current use. Those, plus fine equipment from a great many other manufacturers has kept Sidebanding efficient and enjoyable.

Sideband Gear Today

Today, a number of manufacturers produce CB sideband gear. Some is specialized base station equipment, other rigs are intended for mobile use and still other units may be used for either. And so you should ask, which is the proper set for you, what are the relative values and meanings of these sets, and what are your requirements?

Since there are differences between all sideband rigs, an intelligent purchase will require some thought in advance, especially since you may be heading towards a substantial outlay of dollars.

Before you can decide on which set is for you, take time to consider

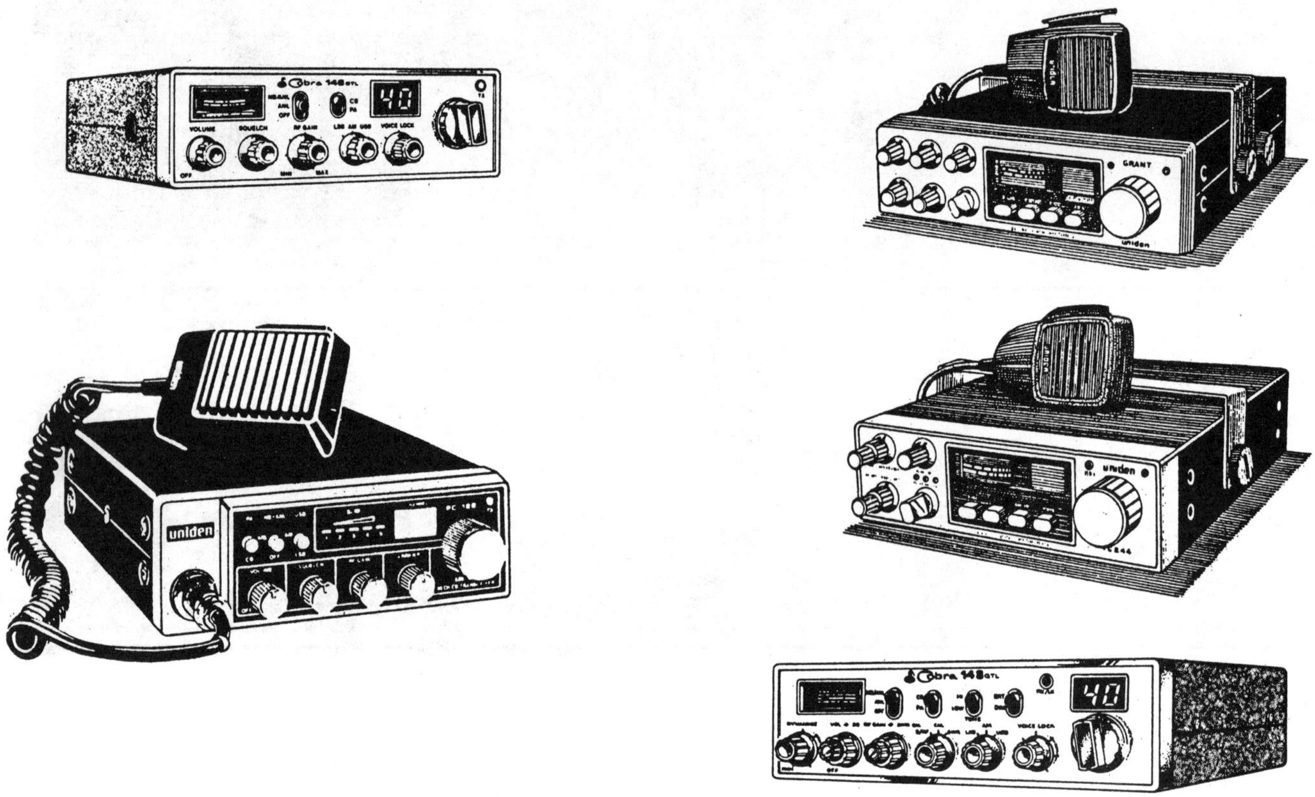

Some popular contemporary SSB/AM mobile rigs: Cobra 146GTL (left top); Uniden PC 122 (left bottom); Uniden Grant (right top); Uniden PC 244 (right middle); Cobra 148GTL (right bottom).

how and where you intend using the set. Only then can you relate your own needs to the selection of equipment available.

So, start off by deciding if you're interested in base or mobile operation, or both. If one set is to be used exclusively as a portable/mobile unit, then remember that mobile-only transceivers usually cost less than those which were designed for base/mobile installations. If, at some later date, you should need to use the mobile rig as a base station, you can always do so by running it from a 6 Ampere regulated power supply at your home.

If a mobile rig is what you're seeking, take the trouble to check out the space in your vehicle where you want to mount the unit. It will have to offer you convenient operating access, yet still give you sufficient knee room. Remember, when it comes to Sidebanding, you will require constant easy-access to the clarifier control, so make certain that it is a prominent feature on the panel so you can locate it without having to take your eyes off the road. Some units are more compact with others, and even "small" sideband rigs are usually bulkier than AM-only rigs, moreover they sometimes contain RF gain and microphone gain controls, in addition to the ever-present USB/LSB selector knob-- and there's got to be sufficient room for such things to exist. The AM-trick of a rig which has all of the controls in the microphone while the rig itself is hidden in the trunk hasn't yet come to the world of Sidebanding.

In fact, in about 1982, the American Antenna Co. (makers of the popular K-40 mobile antennas) designed a very small mobile SSB unit that was smaller than many AM-only units. Being less than 7 inches wide, less than 2 inches high, and 8 inches deep, it turned out to be perhaps too small for mobile Sidebanding. While it transmitted and received very well, the control switches

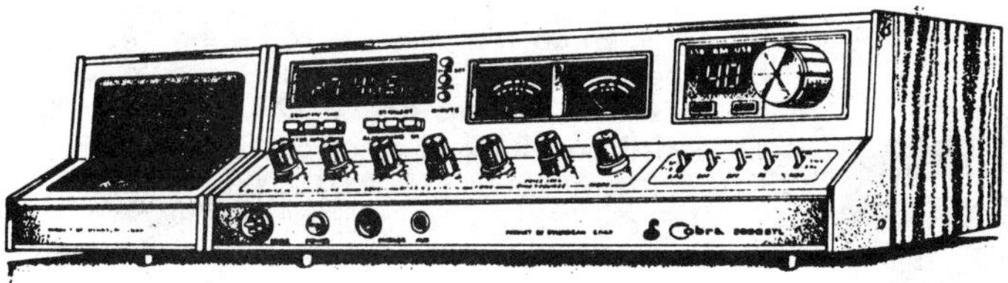

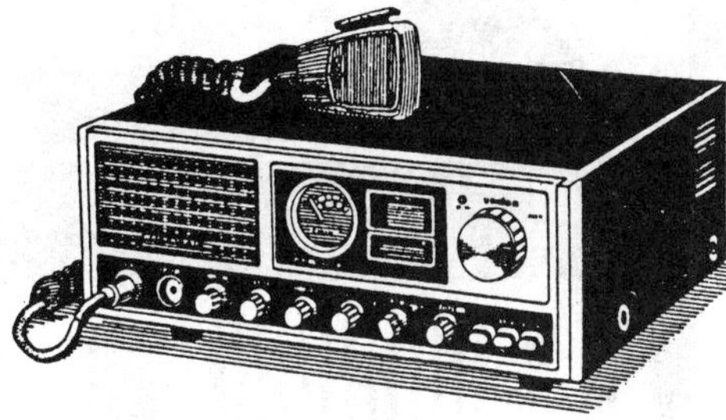

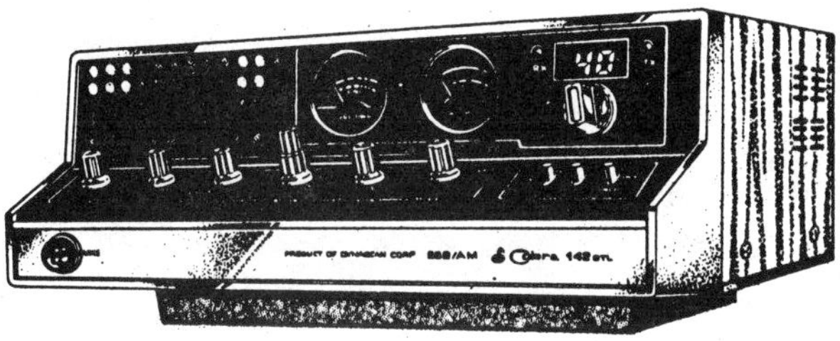

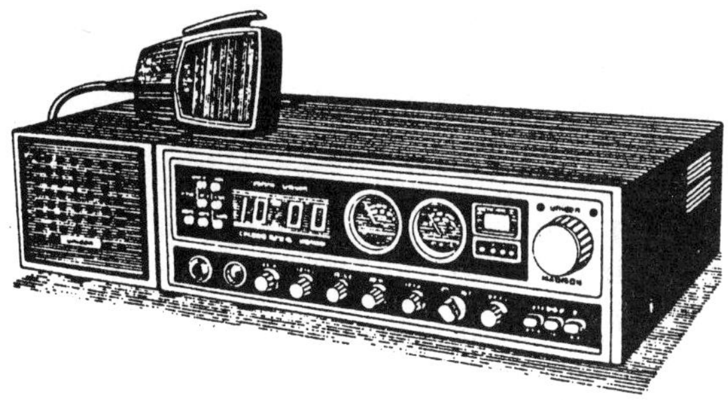

A few excellent modern SSB/AM base stations include: Cobra 2000GTL (left top); Cobra 142GTL (left bottom); Uniden Washington (right top); Uniden Madison (right bottom). The most expensive of these is the Cobra 2000GTL which usually sells in the $370 price range. Shopping around at different dealers can often show wide variations in prices.

and knobs were so small that they were very inconvenient to access while simultaneosly attempting to drive. Ganged controls, having a central knob surrounded by an outer ring control, didn't permit adjustment of one without also adjusting its companion control. As a result, when you attempted to adjust the squelch, you also usually adjusted the RF gain control too; an adjustment of the clarifier invariably meant changing the microphone gain too. This unit was never mass marketed.

Some mobile Sidebanders like to pick rigs that have the clarifier control towards the left side of the panel because they claim it makes it easier to find while driving. Another good feature is a noise blanker, since that takes out or considerably reduces ignition noise. A noise limiter, such as used for reducing/eliminating other types of noise, may not be anywhere nearly as effective against vehicle ignition noise as a noise blanker circuit.

Then, start looking at such things that would make a difference is selecting any CB rig, AM or combined SSB/AM, such as selectivity and sensitivity. Sharp selectivity makes it easier to receive the desired channel while rejecting unwanted transmissions taking place on adjacent channels (known as "splash" in CB jargon). Good sensitivity means that the receiver portion of the transceiver has the ability to pick up weak signals. Both are especially important in mobile installations since mobile whips are less complex than antenna systems than base stations and don't offer the same signal-multiplying gain.

Ask your dealer to show you manufacturers' specification ("spec") sheets for various transceivers you like. These will show the performance capabilities, just the way the spec sheets for air conditioners tell you the cooling abilities and power efficiency ratings, or the spec sheets for vehicles reveal engine size/power and fuel consumption ratings. A simple rule-of-thumb to follow

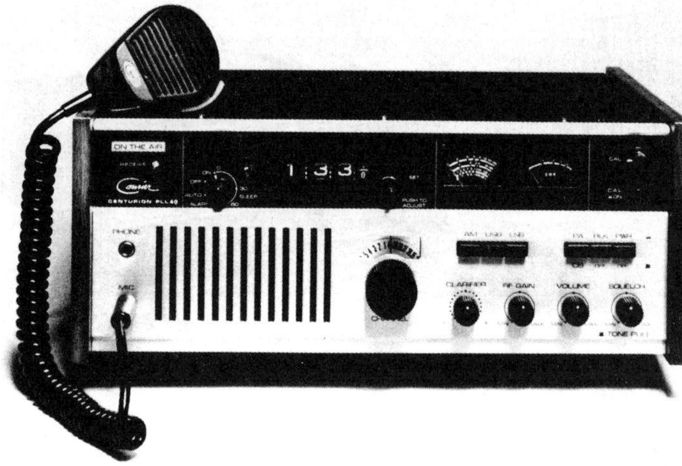

Shown on this and the following two pages are some of the most popular deluxe SSB rigs of the 1970's that are still popular and command top dollar in good condition. At the left is the Fanon/Courier Centurion PLL 40.

The Stoner PRO-40 was designed for SSB-only (no AM) operation. Costing $1,000 new, it had a multitude of exotic features that made it in heavy demand. It was produced in limited quantities and you'll still pay a pretty steep price if you can find someone willing to sell theirs!

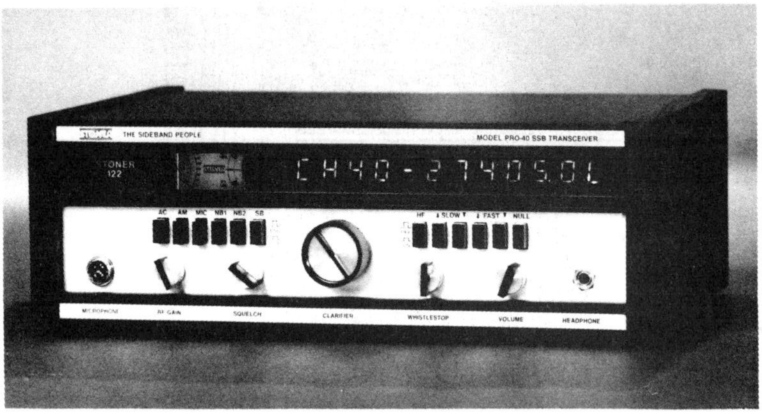

in looking at CB specs is the lower the number used to describe the bandwidth at the highest dB level is the set with the best selectivity. For maximum sensitivity, look for the one having the lowest microvolt (uv.) rating. Nothing is as infuriating as having a receiver with poor sensitivity or selectivity-- or "bad ears," as candidly known in CB-land. To give you some idea of how important good selectivity is, put your CB rig on Channel 18 while you're driving on a busy interstate. Much of what you'll be hearing is the "splash" from all of the nearby cars and trucks operating on Channel 19.

A base station is essentially similar to a mobile unit, except that it has the built-in ability to operate on 110 to 120 volts AC house current. Look for the same basic features with less emphasis on small size. Some of the more fancy base stations offer really nice features including frequency counters, variable tone audio output, external feature accessory jack, headphone jacks, chrome cabinets, S/RF/SWR panel meters, and all sorts of other "bells and whistles" like voice processing, etc.

My own personal philosophy is to get the vesy best rig you can possibly afford-- it reduces the possibility of getting a set that turns out to be less than you wanted or needed for your present or future purposes-- a set that you'll have to suffer with and cuss out and eventually sell to replace with a better unit. A good set is really sort of an investment, anyway. Those who purchased top of the line rigs several years back-- like the Cobra 135, Stoner PRO-40, Browning Golden Eagle Mark IV-- have found that some operators are willing to purchase them in used condition at prices equal to or even more than the original sale prices!

If you can pay the tariff, your best bet is to purchase a set with more features than you feel you might require at the present time. This way,

The Royce Electronics I-640 had some very interesting and unique features.

The Cobra 135 was a dazzling base station that had a mystique of its own and still has a cult of fanatic followers. It's the inspiration for the current Cobra 2000GTL base station, although the old 135's are still sought out by those who love the sets.

as your interest in better operation increases with time and experience, you'll have all of the capabilities right at hand for instant use.

Some CB specialty shops, for a moderate fee, offer to "work on" a new transceiver before delivering same to the customer. Sometimes this can be no more than "peak tuning" all of the transmitter and receiver circuits for maximum functioning, as shown on service bench meters. If the price isn't too high, and it makes you feel better, then, by all means, spring for the offer-- although the actual improvement that you'll get over and above the same rig (in S-units) without these machinations is questionable, or at least open to some debate. If the technician is going to actually modify the unit to "unlock" the clarifier, add "extra channels" or whatever, he should also explain that such modifications are (in the eyes of the FCC) illegal, will permit unauthorized operation, and will also void the manufacturer's warranty. Nevertheless, some operators are happy to get these various extra "capabilities" at whatever the cost. Note also that some of the technicians who perform these modifications aren't as talented as others-- your new piece of equipment has perhaps a 15% to 20% chance of being butchered to one extent or another during the modification surgery.

Note also that some amount of so-called "export" equipment seems to circulate within the CB marketplace. These are transceivers that are ostensibly intended for use in nations other than the U.S.A. and Canada. They may be made to run higher power, operate on more than the 40 CB frequencies authorized in North America, be able to operate in FM mode in addition to AM/SSB, plus do a number of other clever things. They sell at premium prices when they can be located and, of course, their use by American or Canadians is totally outside all of the applicable CB regulations. Recently

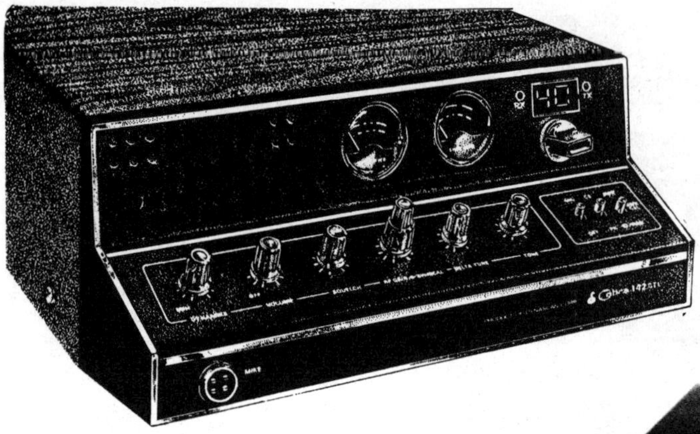

At the left, one of the original "old" Cobra 142GTL's from the 1970's. Although this popular rig is still in production, the original ones continue to be in heavy demand if they remain unmodified.

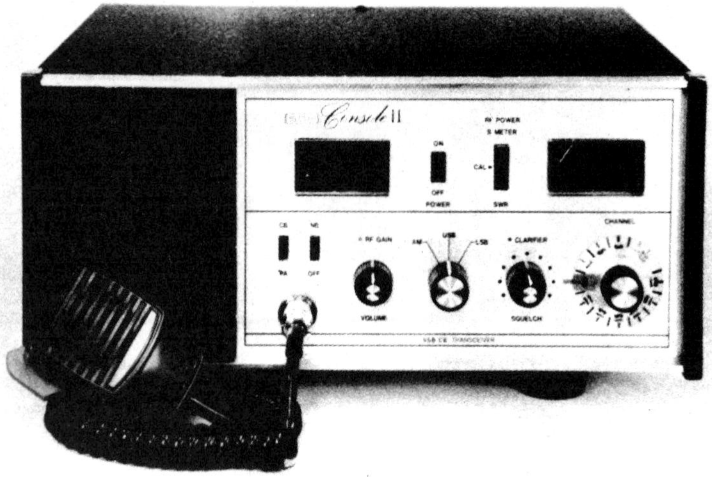

SBE's Console II was a large, solidly-built base station rig that sounded as imposing as it looked.

produced export sets that have created lots of talk include the Galaxy 2100, President Jackson, and Super Star 3900.

These, and even some CB sets that meet FCC/DOC technical standards for CB transceivers, are part of what is often called the "gray market." Gray market sets are produced for sale overseas. Some may be physically identical to standard AM or AM/SSB rigs normally sold in American and Canadian stores and by mail order-- although they possibly are marked with totally different model numbers as their commonly available counterparts. Also, the instruction manuals may well be in languages other than English-- for instance, German, French, Italian, Swedish, and Japanese is common.

These sets aren't intended for sale here, and how they get into the local marketplace is a matter of some conjecture. They often sell at prices considerably lower than comparable equipment made for the America/Canadian markets. Actually, they should work about the same, too. The main problem is that the factory warranties for such equipment are probably not in effect in areas other than those where the sets were intended for use-- like Germany or France, etc.

Another thing to remember is that, as I mentioned, some of this "export" equipment on the gray market is equivalent in every way to garden variety, stock equipment made for the North American markets. Only some of it is the exotic stuff with the "sliders," 26 to 28 MHz extended frequency ranges, etc. I understand that there have been instances where the garden variety CB gear has been misrepresented as being the exotic equipment and sold at outrageous prices to innocents who go off the wall with excitement at the mere hint that they're tasting of forbidden fruit. Be warned!

What About Buying Used Equipment?

The used CB equipment market is booming. Until recently, old CB rigs were sitting in many garages, closets, and basements. These are sets that

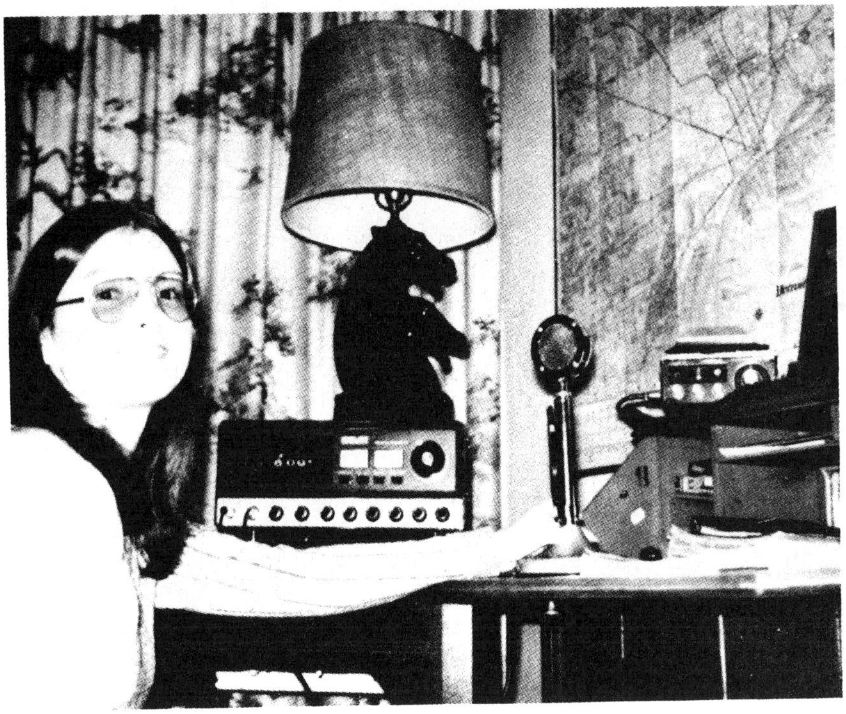

Jill Mc Daris, SSB-2609, proudly displays the Astatic D-104 "lollypop" microphone in use at her fine SSB station.

were probably originally put into service between 1974 and 1981 and then stuck in limbo after the original CB boom years ended. Now that CB has gotten back in gear again, lots of the old equipment is coming out of the closet. Some of it is going back in service to its original owners, the rest of it is being turned into cash at flea markets and swap meets, or traded in to CB shops when newer equipment is purchased.

This equipment is totally different than used cars which seem to be sold and resold over and over again to an endless parade of transactions that come to an end only when the vehicle becomes so decrepit that nobody can figure out how to get it to roll down the street for even one more test drive for a prospective driver.

Many millions of CB sets have been produced over the years, but most of those presently in use don't go back before early-1977. That's when CB shifted from being a 23-channel service to being a 40-channel service. It became illegal to make, sell, or use the old 23-channel rigs. Most of those 23-channel sets probably have an appeal only to those who collect antique CB rigs for their historic value (a growing hobby, by the way).

But the 40-channel rigs are different, and those which were new from 1977 to 1981 or so have been turning up. Of these, some may be in fully operational or serviceable condition and can be purchased at a genuinely good price. But not necessarily so!

The Oldies

Back in the days of 23-channel rigs, if it was a cheap set to start with, it was probably removed from service the first time it had a serious problem that wasn't covered by the warranty. It was replaced with a new rig and either given to the kiddies as a toy or else put away with the thought that someday in the future it might be reapaired cheaply. Today, such rigs are being offered for about $5 on an "as is" basis.

These days, a more expensive CB rig can usually last through its first relatively serious service problem after the warranty expires, and maybe even a few later minor glitches. After that it exists on only the most tenuous level and is sort of looked upon as somewhat of an unreliable traitor to the cause of reliable communications. People will only go so far into pouring repair money into the old set when they could be putting the cash in the general direction of a new rig. Some operators get the thing patched up enough to show it working to unload it on someone else.

The Used Market

By the time a set is ready to be sold off, it's at least a year or more old. Assuming that it's been given normal use, it contains assorted fingerprints, scratches, paint chips, dents, drilled holes, possibly some coffee or soda stains from cups or spills. Maybe it works fine and was owned by a guy who just likes to get a new rig every year. But maybe it's got problems, or no longer has its power cord, mobile mounting brackets, some cabinet screws, etc. So, unless it's a really special rig that happens to be in high demand, like a Cobra 2000GTL or a Cobra 148GTL, then its value to someone as a useful price of communications gear is open to speculation.

Many dealers have a healthy supply of used gear for sale. While many shops will accept the most dilapidated junk for trade-ins, most dealers will throw out the absolute worst sets. Yet there still remain some of poor quality units that really can't be expected to deliver long-term reliable service. Still, these sets may be very cheap and so long as you don't expect too much from them, they may be OK for non-critical applications.

The Tamper Syndrome

Surprisingly few sets last for any more than a year or two without experiencing some amount of tampering or modification in order to improve its operation. This ranges from repainted cabinets to "heavy duty" finals, added channels, souped-up modulation, sliders, and whatever. These modifications may have been of no great service to the cause of good communications.

People who buy used gear fall into several main categories. First, there are those who want to spend the very least amount to get a rig-- like getting a used sideband rig at the price of a new AM-rig. Then there are those who want classic discontinued rigs like a Tram D-201, Stoner PRO-40, Browning Golden Eagle Mark IV, A.R.F. 2001, Communications Power Inc. CP-400, and other equipment of this caliber. Any of these you see at astoundingly low prices are rigs that for one reason or another were rejected by the myriad of fanatics who track down such units with amazing diligence. They might well have been modified and otherwise improved right into oblivion.

How To Be A Sidebander

Having the ability to send out a signal in sideband mode doesn't make you a Sidebander-- no more than owning a piano makes you a musician, owning a wet suit makes you a scuba diver. You have to know what you're doing, the rules of the road.

Anybody who be accepted into the fold and make a go of it in Sidebanding is going to have to come to terms with the basic fact that there's a world of difference between the activities and habits of the Sidebanders than those good folks who operate in AM-mode. A whole new language is going to have to be learned, and lots of operating habits plus a whole old language is going to have to be **un**learned. For some this is problem; for others it's no sweat. There are still others, who, try as they might to sound like dyed-in-the-wool Sidebanders, still let that occasional telltale "10-4" (or worse, "Big 4-10") slip out during a QSO (a QSO is what AM operators call a "shout") to reveal their recent AM ancestry.

Yes, there have been a few operators who flaunted tradition and were determined to perpetuate their old AM-type operating habits with the Sidebanding crowd. At best, they were ignored by Sidebanders. It's been my experience

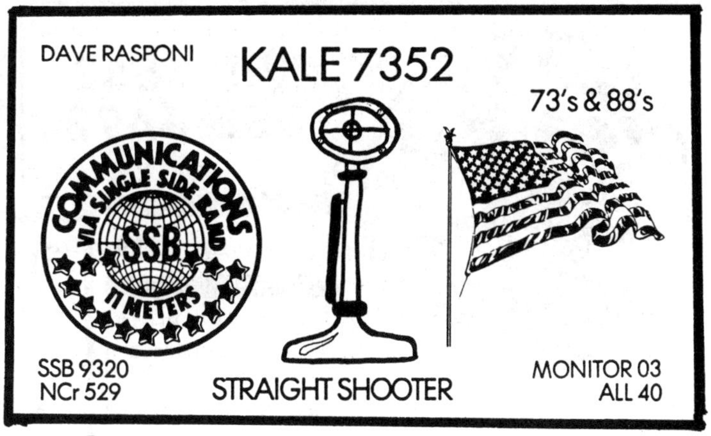

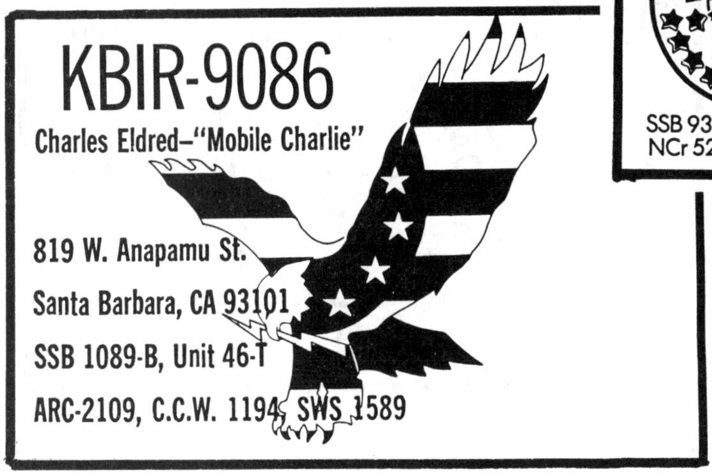

that their usual fate is to become the butt of a considerable amount of scathing, sarcastic ridicule-- Sidebanders have been known to be merciless to those who repeatedly attempt to debase their operating traditions.

Many of the most famous earmarks are unknown on SSB channels. There are some taboos, just as there are many specific requirements-- it's not as freeform as on AM. Violators of some of the taboos (for instance, asking for a time check or 10-36) could well cause an operator to be laughed right out of the county.

Some folks can't hack these customs and rules-- rules that keep SSB something its proponents feel is special that offers the kind of communications the Sidebanders wanted in the first place. If you've read this far into this chapter, we can assume that you're therefore interested in seeing how you can fit into the world of Sidebanding. We'll see to it that you do it without getting razzed off the air.

First Steps

Best bet to start you on the way is to just listen for a day or two to SSB operators in your area without trying to join in. If you can resist the temptation to talk at this time, it will pay off later. There are variations in operating habits in different localities, but you'll note that the one universal truth in Sidebanding is that all operators are expected to show courtesy to other. Rudeness, arguing over the air, and general nastiness isn't tolerated in any form. And stations don't step on one another.

Sidebanding Channels

Technically, all stations have access to all 40 authorized CB channels. But Sidebanders tend to congregate on certain ones, so when you endeavor to monitor these stations it's good to know where they like to operate.

For starters, make note that some Sidebanders use the word "frequencies" instead of the word "channels." Instead of saying "Channel 37," a Sidebander might refer to "375" (Channel 37 is 27.375 MHz). This seems to be a matter of personal preference-- some operators use the channel number designations.

For many years, in the old 23-channel era, Sidebanders preferred channels

15 through 18. Then, when the band was expanded to 40-channels in 1977, many operators moved up to Channels 32 to 40, with some diehards sticking to Channel 16. Generally, on a national basis, most Sidebanders seem to prefer Channels 31 or 32 through 40-- although in suburban and rural areas most stations will tend to operate on only one or two channels in this group.

If you tune to these channels with an AM-only CB rig, you'll hear what sounds like speech that is so badly distorted you can't make out what's being said-- raspy, quacking like a duck. Those are Sidebanders.

Tuning those channels with an SSB rig, you'll be able to copy the stations but only if you've selected the proper sideband-- USB or LSB. Nationally, Sidebanders seem to prefer to operate in LSB mode, although there are local exceptions. For instance, in California, many stations operating above Channel 31 opt for USB-- by mutual agreement. A good idea is to use only that sideband that is preferred in your area by the majority of other operators. No matter what anybody tries to tell you, with present CB equipment technology, local use of both sidebands of the same channel causes a sufficient amount of interference to be annoying.

In some major metro areas, Sidebanders often arbitrarily select one channel for stand-by and calling purposes-- like 36 or 37, for instance. In areas where this in use, the idea is to hold lengthy conversations on some other frequency and not tie-up the stand-by channel. This channel is also a poor place to check your SWR, test your equipment out by whistling into the mike or repeating things like "hello test-- hello test--." Other operators will appreciate your cooperation in leaving this frequency open for calling-- especially portables (that is, mobile units) trying to contact their base stations.

There has also been noted some AM and (especially) Sidebanding use of frequencies 26.995, 27.045, 27.095, 27.145, and 27.195 MHz. These are within the CB band limits, but they aren't authorized for Class D CB communications. These are radio control channels that, long ago, some CB operators realized offered privacy and interference-free operations-- albeit illegal. Since these frequencies lie 10 kHz below CB Channels 4, 8, 12, 16, and 20, they can sometimes be reached by those CB rigs equipped with sliders or other modifications, and those having veriable frequency oscillators (VFO's). Some Sidebanders have also been noted operating 5 kHz below authorized channels-- such as 27.380 MHz-- halfway between Channels 37 and 38. So, just for the record, let me spell out that any operation other than on the 40 authorized channels

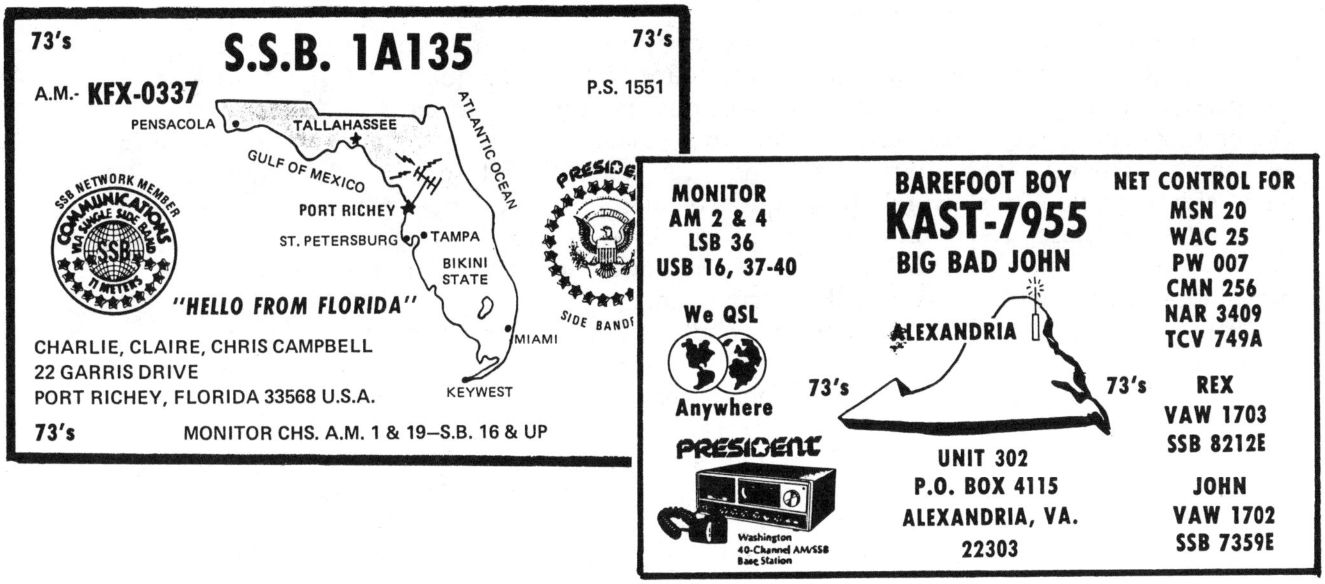

is forbidden and a violation of the FCC's regulations. Operators caught on these off-limits frequencies could be given hassles and fines.

Sometimes operators talk of going "upstairs" or "downstairs." This means shifting above or below the edges of the band-- a practice that is dealt with under the heading "Freeband Radio."

Sideband Identification

One of the first and most obvious things to strike the new listener on the Sidebanding channels is the fact that CB handles aren't used there.

Instead of handles, Sidebanders ID by means of "sideband ID numbers." Such identifications are permitted under current FCC CB regulations. In almost all areas, an operator without numbers is a non-person on the Sideband channels-- and woe be unto the hapless would-be Sidebander who shows up to get acquainted with his new friends and attempts to ID by means of a CB handle. Attempts at operation without a sideband ID number may only mean that you'll be ignored-- but using a CB handle will probably earn you a lecture or some barbs.

In some areas you can be assigned numbers by local and regional groups, clubs, and organizations that issue them as part of their operation. Possibly

---

Why Having A Standardized National SSB Number Is Advisable

"QSK?"

"I heard a QSK, want to join our QSO?"

"You bet, this is QZ-36, first personal John."

"Hi John, welcome to the **Bird Dog** net-- it's our first night. I'm Fred, BD-1-- but I've got a QZ number too, it's QZ-96. Hey, do you know Sam, QZ-25?"

"Is that the same Sam who is SAT-8 in the Saturday morning net on .385?"

"No, but if you're in that net, you probably know Lew, SAT-16-- I believe he was RR-6 or RR-7 when that net was operating a few months ago."

"Can't be RR-7, that was Charlie, over in West Willisville. I just QSO'd him yesterday when we had the **Zulu Victor Net**, he's ZV-6 there."

"How 'bout that! Then, if you're in the **Zulu Victor** group, you must have heard me, ZV-26, Fred."

"You mean, Fred, WOS-46 in the **Wide Open Spaces** group?"

"The one and only! Say, do we know one another? Your voice does sound familiar."

"Freddy, it's John-- WOS-22. Only a little while ago we finished that hour-long QSO with Red and Barney when we were the only four stations in the WOS net tonight."

"Oh yeah-- I knew the voice was familiar. The WOS group sure seems to have fizzled. But, welcome to the **Bird Dog** net-- and we've got Red here on frequency-- BD-2-- so I guess I'll dub you BD-3 in the net. Over to you, BD-3!"

"QSK-- It's Barney, WOS-22, here-- can I join the group, too?"

"Hey, Barney-- you can be BD-4 in our new **Bird Dog** net!"

---

you can get a set at a coffee break or jamboree, and you might even be able to get a set over the air if you can contact someone who has a bloc to give out that will talk to you without having a set in the first place. Depending upon the club activity in your area, you might have do do some searching around from numbers from an established group. This fixation with sideband numbers, which is frequently encountered, need not be a problem since it's possible to obtain a set of national sideband ID numbers quickly, and by mail, from the SSB Network. Check out the section in this book covering clubs-- that should give useful information.

Ultimately, most operators seem to end up with an assortment of ID numbers from various groups. For instance, over the years my own local area has seen a great many local groups come and go. I'm a sociable guy and am always willing to accept numbers offered by these groups. That means that in addition to my national SSB Network number, SSB-13, I'm on the books of various area groups as KW-146, CLI-73, AR-1758, SCS-656, ASB-86, LIS-936 and a score of other numbers. Most of the groups that gave me those numbers no longer exist-- some of them vanished so long ago that most current operators don't even know when they were given out, or the names of the clubs or organizations that fathered them. That's why, since 1964, I've just relied upon my SSB Network ID, SSB-13-- the group is still in existence, and the numbers are instantly recognized by any of the 125,000 affiliated operators throughout the U.S., Canada, and the world.

Access to a Sideband Channel

Although there are variations that are preferred in local areas, some stations simply call "CQ," just as you'd expect to hear on the ham bands. It isn't a very popular approach, but has been slowly growing in popularity. Certainly, the wrong way to access a sideband channel is to ask for a "break," the way so many people do when operating on AM channels. This is perceived as a very bad operating practice.

A typical way of making a call to a known station would be to, first, make certain that the frequency is not in use, then just go ahead and make your call.

Example: "SSB-9344, SSB-9344...SSB-13 calling!"

If the station is listening, you take it from there. If there's no reply, you may wish to release the channel so that others who might also wish to make a call know that they won't be interferring with your attempted call:

Example: "Negative contact...SSB-13 is clear."

If you announce a negative contact but your station is open for calls from others who might want to talk to you, you can close with "clear and standing by on the channel." If you're closing down your station and not open for communications, you say, "clear and QRT." QRT is the code signal for shutting down the station. Stations who announce that they are "QRT and standing by" are displaying their ignorance of operating procedure.

Saying you're "QRT and standing by" makes as much sense as saying, "I'm going out now, so anybody that wants to come to my house and visit is welcome to do so." Clever, no? Yet, you'll hear it very often.

If you don't have any specific station you want to call, but would like to see if any of your pals are monitoring the frequency, all you really need do is key up and ask, "QSK?" That means, something like-- "Hello world-- I'm here, my radio's working, my brain is engaged, my mouth is in gear, and I'm hoping that there's a kindred soul within the sound of my voice!"

You can even use "QSK" to help you hop into someone else's conversation. When there's a gap in the chatter, key up and say, "QSK!"-- it tells the other stations on the frequency, "I'm here and I'd like to join in." You'll

The emblem of the SSB Network is known the world over. It represents the solidarity of purpose and sense of cooperation and friendship that has made Sidebanding what it is today.

usually be invited in with a hearty welcome. Other stations may join in, too and in short order you'll be a part of one of the roundtable discussions for which Sidebanders are so famous. Not long ago I hopped into one of those gabfests that had more than 20 stations taking turns offering opinions on the past, present, and future of Sideband equipment! It was fascinating.

Some Taboos

I've already mentioned that CB handles are definitely out on sideband frequencies, so are all of the AM-side CB buzzwords like "cottonpicker," "mercy, mercy," "roger-D," "10-roger," "4-10," "40-roger," wishing "good numbers" upon others and hoping they'll have "a good day today and a better day tomorrow"-- in fact all of that type of talk is really frowned upon by all concerned. So are time checks and radio checks, as well as "quickie breaks," "shortie breaks," or any other kind of "breaks" or stations identifying themselves as "breakers," or asking anyone to give them a "break."

The liberal use of "fer sure," as popularized on AM, is out. To Sidebanders, the FCC is (as on AM) often called "Uncle Charlie," and you'll also hear Sidebanders talk about getting "stepped on," or "walked all over," or tell other stations with strong signals that the're "walking the dog" or "blowing smoke"-- which you'll also hear in AM operations. But while a Sidebander may refer to a "Smokey," you will never hear a "Smokey report" on a channel used by Sidebanders.

I mentioned that "radio checks" are frowned upon, but that isn't to say that there aren't any exchanges of information between stations on how some new antenna, mike, or rig sounds compared to something else. The expression "radio check" strikes most Sidebanders as being said in the sense that is is used on the AM channels-- as a very thinly disguised ploy for the sole purpose of establishing a superficial quickie contact with a random station. Those who are experiencing a genuine equipment problem or are trying out newly installed equipment are certainly not ignored-- but I think that most who hear "Whadda my hittin' wa with?" view it as a message from a transplanted AM-side operator who seeks to perpetuate that brand of operating practice and bring it into Sidebanding.

If you tell another operator that you think your gear isn't working properly, and explain why, or describe your newly installed rig or antenna-- that's usually sufficient to spur the other station into advising you how you sound. You won't get a "nine pound" or S-meter reading, however. You'll probably get a report that's something like 5-by-9, 3-by-4, or something along those lines. The first numeral describes, on a scale of 1 through 5, the readability of your station. The second digit, on a scale of 1 through 9, gives you a general description of the strength of your signals.

The readability numbers are: 1 = Unreadable; 2 = Barely readable; 3 = Readable with difficulty; 4 = Readable with a little difficulty; 5 = Perfectly readable. Signal strength ratings are: 1 = Barely perceptible; 2 = Very weak; 3 = Weak; 4 = Fair; 5 = Fairly good; 6 = Good; 7 = Moderately strong; 8 = Strong; 9 = Extremely strong.

The reason for the dislike for Smokey breaks, time checks, radio checks, and other such nonsense is, I think, merely expressing a fear that Sidebanding could deteoriate into a scene where the channels are occupied primarily by people looking for very brief, meaningless, shallow hit-and-run contacts that are so totally annoying to operators who pride themselves on their comeraderie and love of conversation. The thought of stations turning up for the sole purpose of occupying a channel for hours asking every station within earshot to tell them if their transmitter is working is their worst nightmare come true!

So, those who normally use AM words, terms, and expressions might do well to become acclimated to the way it's done on the SSB'ing channels, taking careful precautions to avoid any of the more dreaded aspects of AM-type operation. Like, forget that you ever heard that there are such things as "10-Codes," and words like bodacious and negatory. When in doubt, talk on Sideband in plain English-- Sidebanders would much prefer to hear regular English from someone still unfamilar with things such as "Q-Codes," or who elect not to use them-- than all of that AM-side CB lingo.

The Sideband Codes

Although the 10 Codes are ignored by Sidebanders, you'll hear excerpts from the series of international Q-Signals. Sidebanders have selected a few of these signals from this very lengthy set of codes. You may already be familiar with ones such as QSL, QRM, etc.-- and I have already explained how QSK is being used by Sidebanders.

My own personal preference is to use these signals rather sparingly in my own on-the-air contacts. Some seem to fit naturally into a sentence, but others seem to very uncomfortably into a conversation-- like hearing someone show humor on the radio by saying "hi hi" instead of actually laughing they way they would if you were talking to them in person.

I find it particularly irritating when another operator wants to know if I am copying his transmissions or information and keeps asking me, "QSL?"

Most likely I've become a bit jaded about the overuse of these Q-Signals and because so much of the time they are really butchered by those who don't know how to use them and seem to think that you'll think that they are really hotshot radio operators if they bombard you with a bunch of them. If I had only a $1 bill for every station I've heard this past week misuse one of these codes (especially QRT), I'd probably have enough for a new Sideband rig with enough left over for the best steak dinner in town!

Therefore, in this book I've included a listing of the more frequently encountered Q-Signals noted on sideband frequencies-- along with the commonly accepted usages and meanings. These meanings don't always exactly correspond with the usage or meaning found for the same signals in other radio services, such as ham radio. I'll be the first to admit that Sidebanders have added a few twists and shadings of their own.

Most Popular Sidebanders' Q-Signals

Sidebanders don't use the 10-Code, nor do they tolerate its use. Instead, the use and misuse the Q-Signals-- or at least certain ones selected from the many that exist. Here are the ones you're most likely to encounter and how they are used and misused.

QRM This refers to man-made interference such as other stations on the frequency, bleedover. Usually called "Q-R-Mary."

QRN Interference, but not of human derivation-- like static, ignition noise, etc. Often said "Q-R-Nancy."

QRT Means: "I'm leaving the air and my station will be closed. I'm pulling the big switch and not listening for any more calls, so do not attempt to contact me." It's one of the most misused Q-Signals. It does **not** mean that you're standing by for more calls. Every time you hear some operator say that he's "QRT and standing by," you're hearing a person demonstrate their ignorance.

QRX This indicates a brief and deliberate interruption of communications. usually for a stated period of time. QRX-5 means that there will be a 5-minute interruption while you answer the door or 'phone, etc. QRX-1 means "wait a minute."

QRZ Said as "Q-R-Zed." It means, "Who is calling me?" Used when you heard your numbers called but couldn't catch the numbers of the calling station.

QSB Fluctuating, fading, or fluttering signals. Not especially applicable to Sidebanding, but a Q-Signal heard nonetheless.

QSK Technically means: "I can hear you between my signals." To Sidebanders it signifies, "Are there any local stations on the frequency that would like to chat with me?" It's also said when a station wants to let others using the frequency that he's interested in joining in the conversation taking place.

QSL An acknowledgement. If an operator asks "QSL?" he wants you to state that you copied the information just sent. The proper response is "I QSL." A QSL is also a station card.

QSO A "Cue-so" is a 2-way contact or conversation.

QSP Relay of a message. (Not often encountered.)

QSX Means I'm standing by and monitoring, available for any station that's looking for me. Not exactly a solicitation for contacts, just telling other locals that you're tuned in and listening. Usually said at the end of a contact, as appropriate.

QSY Switching to another frequency, usually specified as in "QSY 375" meaning, "switching to 27.375" (Channel 37).

QTH Means location, never given as a street address. A DX station asking for an operator's QTH needs only to know the city/town or state name. An operator wanting a mailing address will usually request an "A.D." (address) including a "zipper" (Zip code). A local station asking for QTH info need only to be told something like, "A mile east of the fire station," "Near the intersection of 4th and Spruce Streets," "Over near the river," or something of that nature.

QTX Used by some operators as QSX is used.

Outrageous Sidebanders' QQ-Code

QQA You've tried USB, you've tried LSB, even both sidebands at once. Hope you're satisfied. Now will you go QRT so we can use the frequency?

QQB Funniest thing about hearing you on SSB is that with only one sideband you sound twice as idiotic as you did on AM.

QQC I love your new slider. Can you slide down about 500 kHz?

QQD ATTENTION! To the AM station here on a Sidebanding frequency. Just

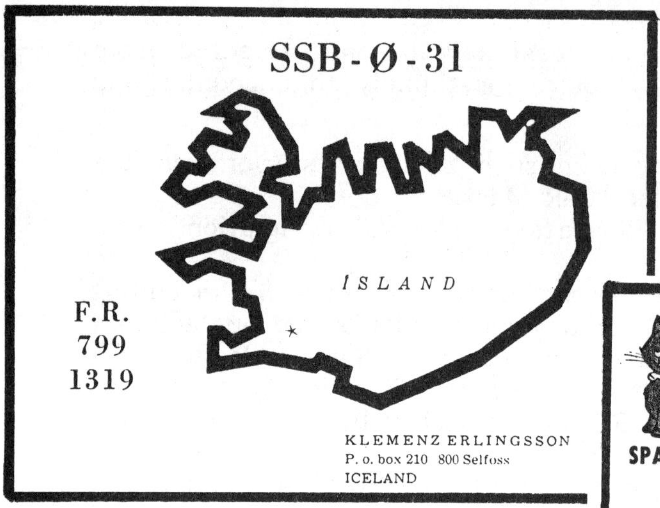

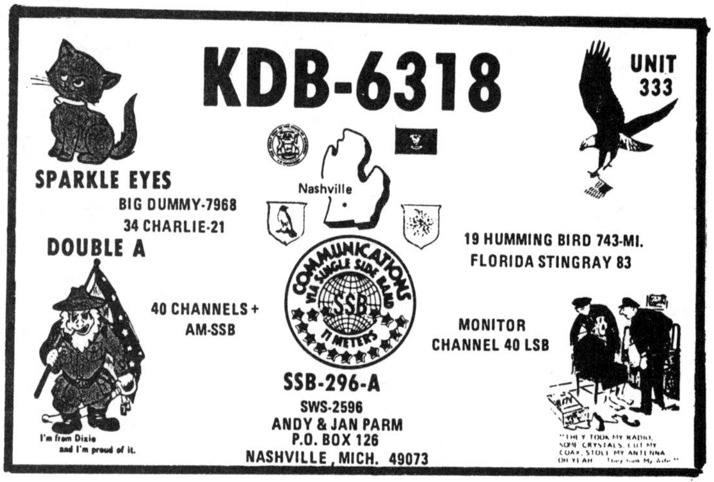

because they won't talk to you on the AM channels, what makes you think we want you here?

QQE HELP! It's the invasion of the 10-4 people!

QQG To the station that keeps asking for a 10-36: The time is right for you to go back to AM.

QQH To the station asking for a "radio check." Sorry, but I closed my radio checking account when I left the AM channels.

QQI Yes, all stations do copy you, and we all agree you're a boob.

QQJ To the station giving the 1-2 radio check numbers. Your signal strength is about 3, your IQ about 1-2.

Q-TIP To the station announcing "QRT & standing by," better use a Q-Tip to clean out your ears and learn the Q-Signals.

A Few More Thoughts

Don't announce your last name, street address, or landline (telephone) number over the air. Don't ask anybody else to give you theirs over the air. Many Sidebanders are also licensed in the Amateur Radio Service and may mention this fact in a conversation. They don't usually care to announce their ham callsign over the 27 MHz channels and would appreciate your not asking them to do so.

Also, some operators have equipment that has been modified or enriched to the point where it bears but scant resemblance to a rig that the FCC would recognize as being usable in the CB service. While an operator might go so far as advise you that the ol' rig had been "worked on" or had "extra freqs" or "capabilities," you'll seldom hear a detailed discussion of what work has actually been done-- nor will you hear any requests for such a description. Some things are just left unsaid, and probably better so!

Be Yourself

Simply because a Sidebanding frequency isn't buzzing away with chatter doesn't mean that there aren't many other operators monitoring there.

There could be dozens or even hundreds of operators within the range of your station. Every time you key up your mike you are telling these people something about yourself-- whether you are a considerate or inconsiderate person-- a good operator or (horrors!) a "lid."

The best asset, I've always felt, to any communications effort isn't necessarily a super rig, a monster antenna, a clear channel, etc. Every time, I'd say that a good operator is the key-- someone who knows how to take any given station and enliven it with an efficient and courteous on-the-air image.

One of the reasons that that there are so few rotten operators in Sidebanding is that other operators don't want them there. No, they don't have any magical spells to keep lids away, they just won't talk to any. Sure, they could net trapped once or twice by one of these people, but with so many stations listening in, a poor operator soon racks up a very undesirable reputation. With nobody to talk to, these people eventually drift off into other amusements. The silent treatment is an especially potent way of keeping the frequencies clean and sane.

So, be yourself. Don't come on strong in an effort to impress Sidebanders with a lot of hullabaloo. If you're new at the game, don't be afraid to ask some questions about local operating practices and customs; admit that you're a newcomer. They'll probably welcome you with open arms and offer you every encouragement-- then watch you for a while until they get those good vibes that let them know that you're really one of their own.

Using Sideband Gear

Once you've gotten that beautiful sideband rig into your vehicle or base station, your antenna is ready to go, and you understand the operating techniques used in your area, you're still faced with operating the equipment. AM gear is so straightforward that it's as basic to operate as a TV set or transistor portable AM/FM receiver, but sideband gear offers a few challenges.

Actually, all CB equipment is designed to be operated with a minimum of grief, you shouldn't face any of it with trepidation; however, there are a few tips that don't seem to get included in the operating manuals packed in with the equipment.

Even though the books don't say anything about it, and many experts have told me its total nonsense and a throwback to the old days of vacuum tube equipment, I like to let the rig "warm up" for 15 to 20 minutes before I start talking. It's not really a must, but I've found that they seem to work better after they've sat there cooking for a little while. Probably all in my imagination or simply a nostalgic feeling for the old days, or maybe there's something mystical about it-- but I never get those glowing reports on how great the rig sounds until it's fully awake.

Well, be that as it may, when you're ready to begin operating, find the center slot of your selected frequency by using the clarifier control. Do this slowly and carefully. Rock it back and forth gently over a signal being received. Notice how the operator's voice goes through several stages sounding like a robot or Alvin the Chipmunk. Somewhere, however, within the adjustment process, you'll find one single exact spot where the other station sounds reasonably human, or at least humanoid. Actually, sideband equipment and techniques have progressed to the point where, on many of the better pieces of gear, the incoming voices sound quite good. On most of the early SSB gear you were never ever able to achieve results better than odd-sounding, interplanetary-type mechanical voices.

If it turns out that, try as you may, you're unable to get the other fellow clarified to the point where you can make any sense out of what he's saying, check your USB/LSB selector knob-- he may be transmitting on one sideband while you're attempting to clarify him on the other sideband-- a trick that doesn't work. While it's generally considered poor operating practice for stations to transmit on the normally unused sideband of a channel in a local area, some operators still insist on doing it anyway. On the other hand, he may may be on the proper sideband, and your selector switch may be the one that's out of kilter. On 27 MHz, use of LSB predominates in most areas, but there are many local area exceptions.

Stations that come through on your receiver accompanied by a howl are probably itinerant AM operators coming through your sideband receiver. Stations that sound like Donald Duck are Sidebanders all right, but you forgot to switch your rig over to either LSB or USB.

One further thought on stations that can't be clarified-- they may be operating off center slot to the extent that they are past the point where your clarifier can tune effectively. This could happen when someone with a slider moves 1 or 2 kHz off frequency, a poor habit shown by some operators.

Once you find center slot, wait until it's clear of local transmissions before you attempt to transmit. If you step on another local, you'll probably bring upon yourself some flak from the victims of your indiscretion who will no doubt lecture you on the subject of courtesy. Because of the extended groundwave range of sideband, it may be that you do hear some distant stations. Once you determine that they aren't right in your primary coverage area, you can assume its safe to go ahead with your transmission. If you're jamming them, they'll let you know. If they're really far away, you can ask them to back down slightly on their audio (mike gain) or aim their beams away from your area because they're booming in a long way from home.

When you decide to transmit, keep in mind that the output of your sideband transmitter can be adjusted by your variable microphone gain (if your rig has one). For local contacts, back down on the mike gain to the lowest point required for suitable communications. For long-hauls, or when interference is bad, open up the audio to full blast. Always transmit long enough for other stations to clarify you in on their equipment. The AM practice of brief 1 or 2 word transmissions isn't good practice on sideband, not only because it is the trademark of an operator who can't think in complete thoughts, but because it's not enough to permit clarification.

Sideband operators who used to be AM operators are prone to telling others that they are "overmodulating" when they hear distorted SSB audio. While overmodulating is a perfectly good description for what happens when someone attempts to push too much audio (mike gain) into an AM transmitter, on sideband the result of the same practice is more appropriately called "flat-topping." It means that a station is trying to pump more audio into the transmitter than the unit can deal with, and the set just isn't up to the demands being made upon it. The rig does as much as it can to serve you, within the limits of its 12 watt PEP design-- but the modulation-regulating circuits in the transmitter will slice-off the voice peaks if you attempt to run your power mike gain up too high, or yell into the mike. The station that's monitoring you will hear a distorted rendition of your voice in either case.

If you're in contact with a local station that sounds distorted, before you suggest that the there's a flat-topping problem, try backing down on your receiver's RF gain control. It may be a simple matter of the other guy being so close to you and/or so strong that the front end of your receiver is overloading. That could cause the distortion. By backing down on your RF gain control, you cut the incoming signal down by a few pegs, and also reduce a lot of background noise. But don't forget to return it to its maximum gain point (full clockwise) when that contact has ended.

The RF gain control affects only the receiver, not the transmitter output.

In the general operation of your station, if another operator with whom

you normally communicate suddenly tells you that you don't sound as good as you used to (unless you've got a cold or laryngitis), assume that there's a problem at your end. You might come to this conclusion if you aren't reaching your friends on the other side of town the way you did the week before. Maybe you've got an antenna problem. If you don't have an SWR bridge then at least make a cursury eyeball inspection of the antenna, cable and connections. Do this anyway whenever there's been severe weather (snow, ice, wind, lightning).

Sideband— Organized?

CB operators of AM and Sidebanding persuasions have long gotten together in clubs, groups, networks and other organizations. The earliest such group that I can recall was the famous Five Watt Wizards, headquartered in the borough of Queens in New York City. It had its organizational meeting in November, 1959! This was, of course, an AM group since sidebanding hadn't yet made its appearance on 27 MHz. The club lasted for several years and by the time it ran out of steam, there were many other groups going throughout the CB world.

AM and SSB organizations come and go, some are worthwhile and others can't seem to get off the ground. Because Sidebanding is made up of operators who are generally in long-term contact with one another, rather than on AM where hit-and-run quickie contacts between strangers is popular, organizations have been an integral part of the scene.

Some groups are established along very formal lines with elections, regular meetings and dues. Other groups on the Sidebanding scene bear so lightly upon the time, the personal lives, funds, and activities of their members that they're hardly any bother at all.

Then there are those that have come along that appear to be mostly interested in offering group insurance policies, equipment sales, group vacation packages, and all sorts of other promotions.

There are and have been countless local sideband groups-- sometimes several in a single area-- that are well worth the support of operators. If there isn't yet one in your area, you might think about putting one together. Limit membership to active Sidebanders and center the motives around the

Charles ("Chuck") Napier, SSB Network member SSB-55. When not modulating, Chuck's a busy guy in Hollywood. He's appeared in numerous TV series and films, including Rolling Thunder, Handle With Care (also known as Citizens Band), and Rambo II, among many. You just never know who you're liable to meet up with on a Sidebanding frequency!

common interests of the members. Add some public service, a common desire to maintain the integrity of local Sidebanding frequencies, and you've got it made!

Meetings don't need to have all of the parmiamentary claptrap and reading of the minutes of the last meeting, just some coffee and doughnuts and possibly a guest lecturer to demonstrate some new equipment or discuss some technical aspect of communication. Community service could include handling communications during local parades, coffee breaks to raide money for a worthwhile charity or to help out some area resident who has had an especially bad time of things. You could work with public safety agencies in search/rescue operations, work with lonely residents of a rest home or a VA hospital, participate in a neighborhood anti-crime patrol, etc.

Through it all, get plenty of publicity from your local media. Wouldn't it be great publicity for area residents to find out that there's such a thing as the Bishop County Sidebanders, the West Taylorsville Sideband Club, the Hunt Lake Sideband Association (or whatever)? Take photos or videotapes of your activities and circulate them to the media along with press releases.

On a national basis, the oldest and longest active national Sidebanding organization is the SSB Network. It's not a club, it doesn't offer insurance policies, and there aren't any yearly dues. Yet, it's gained a reputation for being the one vital element that binds together Sidebanders throughout the nation and the world into one cohesive group-- regardless of local club affiliation (or no local club affiliation) by its affiliated operators. It's a network, that is, a gathering of operators sharing an interest in Sidebanding, who are in contact with one another. Affiliation is open to old timers, newcomers, and even those who are planning on becoming Sidebanders.

Besides the social aspects of network affiliation, members of the SSB Network have been vital communications links in natural disasters and major accidents. It's the only 27 MHz organization ever to have been mentioned in the Encyclopedia Americana.

Members come from all walks of life, and include a number of notables. For instance, Gil Kowols, the "father" of 11-meter sideband equipment design, was given the SSB Network sideband ID number SSB-27. Another member is rock music star Gary U.S. Bonds, also known on sideband as SSB-17. Another entertainment personality is Charles ("Chuck") Napier, SSB-55-- you've probably seen him in Kojak, the Rockford Files, the movie Handle With Care, and many other TV programs and motion pictures.

The SSB Network began back in 1964 as an informal group of Sidebanders who all had Mark Sidewinder SSB-27 rigs. The operators would gather for regular on-the-air chats in the New York metropolitan area. To add a group

SSB Network members can get this large gold and black membership certificate handsomely inscribed (in calligraphy) with their name, SSB Network ID number, and the date of their affiliation.

identity to the fact that, as Sidebanders, they were a breed apart-- in their approach to communications as well as from a technical standpoint-- all network affiliated stations were given numbers commencing with the letters SSB. The operator who came up with this idea was the original Net Control Station, Russ Walters-- he became SSB-1. I became SSB-13. Thus was the SSB Network was born.

Eventually, operators in surrounding counties began getting into Sidebanding and asking how they could affiliate and it soon became apparent that the concept was something that would become wide-ranging in scope. As new sideband equipment arrived on the scene, the SSB Network's influence and affiliated station fanned out from state-to-state, into Canada, and then across the oceans. Among the many things the SSB Network has accomplished for the benefit of the Sidebanding community was the concept that when the FCC the 17 additional CB channels were allocated to CB in 1977, several of these frequencies should be staked out for exclusive Sidebanding operation.

The SSB Network's affiliates all receive "national" numbers. That is, sideband ID numbers that are appropriate for local communications, and also when you're off and about just about anywhere! Unlike local numbers which are usually unknown only a few hundred miles from their source-- and all too soon may become orphans when their originating organization becomes defunct-- SSB Network numbers are recognized everywhere and identify their user as a member of the original Sidebanding network that has been continuously active since 1964, and now has well over 125,000 affiliates. That means, no matter where you operate-- at home or away-- you aren't a stranger!

SSB Network numbers are issued only by mail. They ask that those seeking affiliation send them a stamped, self-addressed envelope for an application. There is a one-time nominal registration/processing fee for permanent affiliation. The application should be requested from The SSB Network, P.O. Box 908-X, Smithtown, NY 11787. No many how many other sideband ID numbers a person may pick up along the way from local, county, regional, or other types of sideband organizations, national numbers are the ones that identify you as an integral part of the national Sidebanding movement. Many local organizations don't even have their own numbers, requiring all of their members to use only national SSB Network identifications.

Membership materials include a wallet-card bearing the members personal SSB Network ID number, name, and date of affiliation. There's also a handsome gold/black wall certificate with this same information professionally done by hand calligraphy. Included is additional assorted membership informational

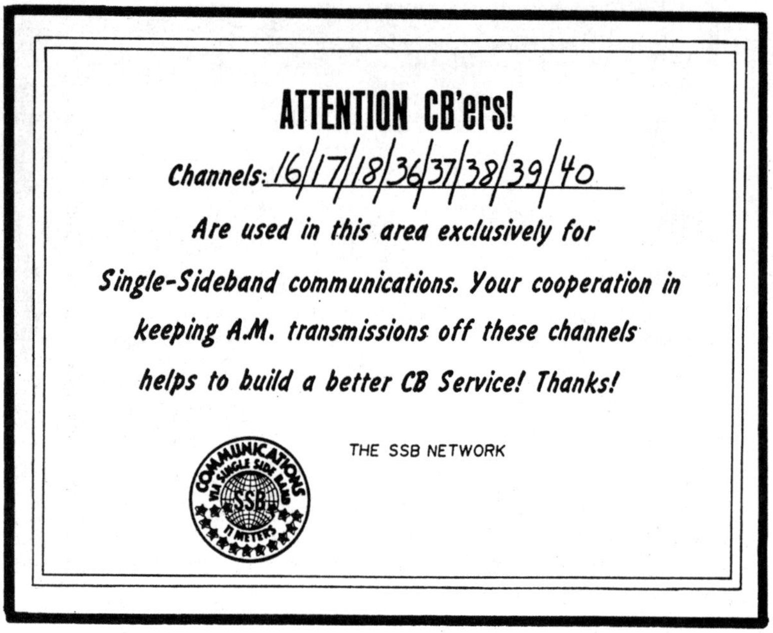

In the 1970's, when AM interference on Sidebanding frequencies was at a peak, the SSB Natwork designed these posters that could be displayed at CB shops and jamborees to educate AM operators about staying away from frequencies normally used by Sidebanders.

literature as explained on the application.

In many areas, there are weekly informal SSB Network rap-sessons organized by local operators. Although the details vary from area-to-area, many times this takes place at from 7 to 8 PM local time on Channel 38-LSB. If no such on-the-air gathering exists in your area, you are welcome to round up some other SSB Network members and get the ball rolling. Also note that several local groups of SSB Network affiliated stations have established emergency monitoring stations on Channel 9-LSB.

AM Interference

One of the more irritating interference problems Sidebanders face from time to time is when AM stations decide to operate on the frequencies used in your area for sidebanding. These stations may be calling for a "break," or can appear in a group to continue a conversation that they had begun on on some other all-AM channel.

If the AM stations are right smack on center slot and your receiver is zero-beat with the carriers there will be few problems. In most cases, the stations are sufficiently off center slot to cause an annoying heterodyne howl to pour forth from your sideband rig. If the AM'er is close by, it can really wipe out the frequency.

While it's true that CB stations, insofar that the FCC is concerned, are able to use any and all of the 40 channels regardless of AM or SSB persuasion, best communications for all concerned take place when the AM and CB stations don't try to integrate on the same local channels. Sidebanders seldom stray onto local AM channels, the scenario is usually AM stations on the frequencies normally used by Sidebanders-- of course, the fact that there are many more AM'ers than Sidebanders in the world gives them some strength in numbers. Then, Sidebanders have never asked for the use of more channels than their share-- actually only a paltry few of the available 40. I've found that an AM operator who shows up on a local Sidebanding channel usually fits into one of several possible categories.

Most commonly, it's a new or unseasoned AM operator who happened onto the frequency by sheer accident in the process of trying for a break or radio check on each of the 40 channels. A couple of shouts are made on the channel and there is complete puzzlement as to why nobody came

The FCC has long been on the receiving end of letters, pleas, suggestions, and formal petitions intended to convince that agency of the need for certain channels to be designated for the exclusive use of Sidebanders. The FCC has always ignored such requests.

back. Many Sidebanders ignore such stations hoping the operator will eventually get tired and go away. This is a poor approach because he'll probably be back again. Wouldn't it have been better to advise the operator of the fact that he is on a channel normally used only for communications in sideband mode? This can be done courteously, and local Sidebanders will usually tolerate one of their ilk briefly switching over to AM on the frequency in order to deliver the message. If you feel that the explanation will take some time, or the other operator has some questions, you can always move over to a more appropriate AM channel for the rap session. This way, he won't keep shouting for a break in AM on the Sidebanding frequency and, who knows, you may well make a new friend for Sidebanding who could eventually decide to join your clan and become one himself!

Then there are those who know full well that they are putting their AM carriers onto a local Sidebanding channel, and either don't care or are going out of their way to deliberately be a nuisance. Maybe they were looking for an empty channel for their shout and one of the area Sidebanding channels wasn't in use at the moment. Experience has shown that these are exasperating and frustrating situations to deal with, and they are especially alarming to Sidebanders. This is because there is always the fear that if enough AM'ers decide that dormant Sidebanding channels in the area are clear spots for their AM operations, then there's real possibility that they could be lost-- it's known as "the invasion of the 10-4 people."

You might try switching over to AM-side in order to explain that AM operators normally use the vast majority of channels in the band and that all Sidebanders ask is to be left alone on a few local frequencies-- that it's for everybody's mutual benefit because AM operators wouldn't like it if Sidebanders suddenly began randomly showing up all over the band. This approach may work, but it's a longshot. Since this problem only happens on unused channels staked out by Sidebanders, the trick is for Sidebanders to use those channels-- as the SSB Network said long ago, "use 'em or lose 'em." Frankly, you can't blame an AM operator for getting his shout through on your quiet and unused channel when all of his are crowded.

It's a bad practice to blow your cool with AM interlopers on Sidebanding frequencies, just as it is to seek retaliation in kind by moving your sideband operations to AM frequencies. This only makes matters worse.

Special Mobile Sidebanding Thoughts

Sidebanders in portable (mobile) units should take into account that they won't be receiving road condition reports, Smokey reports, and other normal Channel 19 AM-side type communications. Those who need that type of information are going to have to switch to Channel 19 or other locally used AM mobile enroute channel for such advisories.

It's probably also not very practical to try and handle a road emergency via sideband on Channel 9, unless your area has it's own Channel 9 sideband monitor. Most regular Channel 9 monitors are set up for AM operation only.

The evening rush hours in urban/suburban areas seem to the the highest activity time for on-the-move Sidebanders, many of whom are trying to get through to their own bases. This is a good time to keep your own contacts as brief as possible so as not to tie up the frequencies so those portables can get through. Portables don't have the punch to get through a gaggle of base stations occupying a frequency with a marathon talk session. It's especially advisable for base stations to allow a short two or three second lag between transmissions so that a calling mobile can make itself known. Courtesy calls for mobiles to be given preferential use of the frequency-- and if a mobile unit is having a problem getting through to its base, it's always good policy to offer to be the go-between for a relay.

Extracting Yourself

Backing out of a sideband contact is rather straightforward. You simply tell the other station that you've got to pull the switch-- you may wish to go into more detail, the wife (XYL) or husband (OM) is calling, there's a telephone call, you've got to paint the porch, there's black smoke and the odor of burning insulation coming from the rig-- whatever is appropriate.

Avoid any temptation to wish "lots of good numbers" or "3's and 8's" upon the other station. Nor should you bestow "a good day today and a better day tomorrow," or the hope that the "Good Lord take a liking" to him. Just say goodbye, announce your ID numbers, and leave without any fuss.

SSB IS STILL YOUR BEST BET WHEN IT COMES TO ENJOYMENT & USEFULNESS!

8

Buying Guide for Used Stuff

Here are guidelines for buying used CB equipment:

1. The equipment may well be really good gear that works just fine and is being replaced by newer and more sophisticated gear. But remember that this isn't always the case, it may have some chronic service problem.

2. Buy used gear from a reputable communications shop if at all possible. Such dealers hope that you will continue to do business with them and are least likely to stick you with a clunker. Avoid buying from guys wandering around at a CB jamboree or coffee break trying to peddle a single rig "because I need the money for my daughter's eye (or kidney) operation."

3. Be immediately suspicious of used CB rigs offered minus their power cords. Not that you can't replace the power cord, but there's a reasonably good chance that the rig has been stolen.

4. Does the seller offer any warranty on the gear? No standard practice has yet been established so you'll have to play it by ear. Rigs sold "as is" means that the seller makes no representations as to whether or not it works and doesn't even say for sure that it can be repaired at any price. Try for a 90-day parts warranty, or a 30-day warranty on parts and labor. Naturally, only a dealer could offer such a warranty, and be sure to get it in writing.

So You Want To Buy A Used CB Rig?

READERS' GUIDE TO USED CB EQUIPMENT

1. This little gem was the one with the best signal on the air back in October of 1958. Seen a little wear and tear but still a winner!

2. This one is sold "as is," a piece of wire here, a resistor there, and you're on the air.

3. Our technicians have tested the output on this rig and have certified that it stays within the authorized 5-watt input limit.

4. Built at home with tender loving care, this unique customized rig incorporates multiple frequency doppler phase shift quadruple modulation and a super-hooperdyne receiver.

Countless el-cheapo off-brand or no-brand CB's have turned up on the new/used market over the years. While these sets (usually sold new at discount, novelty, and auto supply stores) are a waste of money in new condition. In used condition, they're worse than a wart on your nose.

The most oddball warranty I ever heard of was called a "lot line warranty." That meant that the unit was under warranty until the buyer crossed the line at the end of the seller's parking lot!

5. Although you should fully expect that the seller will let you try the set at his place, see if you can get him to agree to let you take the rig home for a day or two attached to your own antenna-- with written assurance that you are allowed to return it for full refund/credit if it doesn't work as represented. I know of one dealer who admitted to me that he ran all of his used equipment through a 100 watt footwarmer (illegal power booster) during used equipment demonstrations. At the very least, if you can test the set only in the store, get some on-the-air signal reports, ask to see what the rig's output runs on a wattmeter and modulation meter. Anybody who buys a used CB rig without seeing it work probably deserves what he gets.

6. Can and will the seller service the set after his warranty expires? If not, look elsewhere.

7. Don't be dazzled by the word "reconditioned." Ideally, it would mean that the set has been given a top-to-bottom once-over by a qualified service technician, an authorized factory service center, or the original manufacturer. All circuits, components, and chips were checked as OK or replaced/repaired, or peaked to original specs. The word "reconditioned" may mean only that a damp rag dipped in Mr. Clean was run across the faceplate and cabinet. Get a complete description of the reconditioning (in writing, if possible).

8. Unless you're looking for a rig that's spectacular from a company no longer in business, or in business but no longer making CB rigs, or you're looking at a CB rig that is no longer in current production, you've got something else to consider. Many companies have come and gone over the years, and even within the framework of those manufacturers who made 40-channel rigs only a few years ago, there could be problems. All electronics components arent standard stock items that can be pulled off the shelf to replace the ailing innards of your rig. Many components-- such as switches, chips, IF cans, etc.-- were specially designed and manufactured for particular companies or CB transceivers. Once that company has gone out of business, or the CB business, or the set has gone out of current production, then those special components are no longer produced. Whatever remains in inventory may

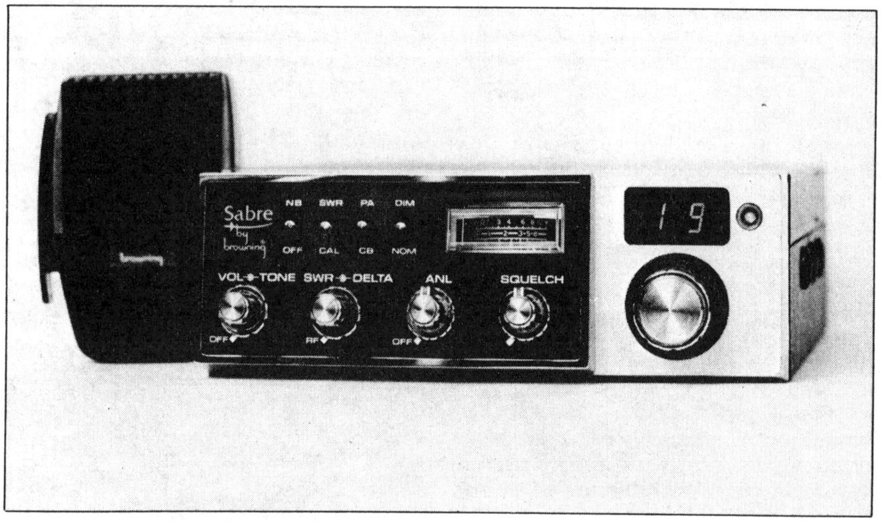

Some companies that aren't in business any longer produced sets that are still in high demand because of the quality of the equipment. Just about anything made by Browning Labs easily fits into this category.

be used for repair purposes, but when they are used up, then the equipment may become a total orphan that can't be fixed at any cost. Even if the set is working now, a repair that it needs in a few months or a year could be a dead end. Be sure that, whatever set you get, comes with an instruction and/or service manual-- hopefully a schematic too. If there's no paperwork with the set, you might be able to use that point as a price wedge-- or to think that it's missing because the rig was stolen.

9. When you check out the rig on-the-air, see how the receiver sounds. Scratchy sounding audio might mean a torn speaker cone or a problem in the audio circuitry. If rotating the volume control causes a crackle noise in the speaker or makes the sound cut in and out, it could be dirty or defective. Does the squelch work? What about the meters?

10. Have someone several miles away give you a report on your signal strength and the quality of the modulation. If the set has sideband, check that out too. Make sure that other Sidebanders can be clarified in adequately. Do the on-the-air tests with someone known to you, Sidebanders normally balk at giving "radio checks" and all too many AM'ers will give you a "9 pound" and "crystal clear modulation" report to anybody who asks, no matter how lousy the signal. Also, turn the channel selector to every single channel and make certain that the set seems to be operational there-- isolated dead channels could mean problems with the channel selector switch or in the phase lock loop (PLL) circuit.

11. Keep in mind that a used rig is something that someone else didn't want for one reason or another. It will never look or work as well as it did the day it was new-- that's why it costs less than it did at that time. So, at best expect at least some minor flaws-- if you want something flawless, spotless, and pristine then buy new equipment. The best you can hope to to is pay a reasonable price, based upon the set's age and condition.

12. Always remember that the price asked for any used equipment is exactly that-- an asking price. Used gear sells for whatever the seller can get from the buyer-- there are no set prices. Make up your mind in advance how much you are willing to spend, and then don't go over that price by as much as a dime. Chances are that the seller's first asking price is 25% more than he'll eventually take for the rig if you are willing to haggle a bit. No matter what he asks, your duty is to look shocked, annoyed, or disappointed because the unit isn't even worth half his asking price-- your friend Phil got one in better condition only last week for much less. Point out all of the set's failings-- its grubby condition, missing power cord and knob, lost operator's manual, needed repairs, etc. From that point, figure that

you should probably be able to get the set for 65% to 75% of whatever the original asking price was-- especially if you take the cash out of your pocket and lay it out on the counter, fanned out like the winning hand in a poker game. Whatever you do, never pay the original asking price, even if its less than you were willing to spend and the rig is an absolute cream puff. Don't let the seller see your eyes pop out in delight when he tosses you a lowball price. If necessary, tell him you need a day to think it over, and leave a small (about $5) good faith ("earnest money") deposit-- but should you decide not to buy the rig, you probably won't get your money back.

Some folks claim that it pays to shop for used equipment with a friend tagging along. They say that it spooks out the seller when it looks like it's two-against-one. And your friend can help in the haggling by agreeing that the price is too high.

When you're listening to the claims being made for the rig, remember that if it sounds too good to be true, then it probably is too good to be true. Also, sales claims and promises cannot possibly be any more reliable than the person making those claims. Do you know anything about the person trying to sell you the rig? Will he give you a written bill of sale containing his name and address, the serial number of the transceiver, the price, plus all of the agreed upon terms of any warranty?

When it comes to antennas, cables, meters, test gear, microphones, and other accessories, stay away from used offerings and start with new gear.

I would really suggest buying a used CB transceiver when it simply isn't

While old pre-1977 (23 or fewer channels) rigs can still be used by their original owners, FCC regulations don't permit them to be sold in used condition to others. Dealers can accept them as trade-ins, but can't resell them anymore. This LaSalle rig is a grisled old veteran of the early QRM-wars and richly deserves to be retired to a museum or a CB collector's shelf.

possible for you to buy a new one, or if your father, brother, or son offers you one in mint condition at a good price. Ultimately, the question you must ask yourself is, "Should I buy a new rig at the same price I can buy an older used rig which once sold for more money?" That's the same as deciding if this year's Chevy is a better deal than a five year old Caddy at the same price. It's a matter of personal preference. Some folks would rather have a new Chevy, Ford or anything in preference to something used, previously owned, rebuilt, or reconditioned.

Sidebanders, in particular, are very much into "horse trading" their old equipment. This is probably so because their gear is so much more expensive than the equipment used on the AM channels. I've heard some Sidebanders sell off the rig they're talking on to another operator they've just met on the air! Undoubtedly this system is satisfactory to some people because it isn't all that uncommon. But buying a CB rig of any kind from someone totally unknown does have its risks.

A fellow I know made the mistake of buying a beautiful sideband base station from someone he met and spent the afternoon with at a CB coffee break. The rig was a popular model, and touted as being "still in its original factory carton" (I've always wondered why a factory carton counts for so many points when buying/selling electronics gear). He took the set out of the carton and looked at it right at the coffee break. It was a dream come true-- a deluxe $500 base station for only $200! The deal was finalized as soon as the spectacular rig was eyeballed-- but when it was taken to its new home and didn't work (not even a little bit), it was decided to check beneath the black satin, chrome, and brushed aluminum cabinet. The "rig" had no guts, only a brick to give it some weight. It was a hollow shell worth about $10-- possibly made to be used as a display-- but it did have that everpopular "original factory carton."

Whether buying new or new equipment, your best bet is an electronics or CB specialty ship, and don't forget that there are plenty of independent dealers as well major chains like Radio Shack from which to select. Be a bit flexible within your general requirements. Just because you are shopping specifically for sets A, B, and C, don't be closed to sets D, E, and F just because you hadn't known enough about them at the start. You may find, as you shop around, that they're more suited to your needs than your original "target" sets.

Ask lots of questions-- ask the same questions in each store; look at spec sheets. You'll probably find that the personnel in specialty shops are invariably very well-informed, helpful, and anxious to answer your questions. Be aware, however, that they are anxious to have you for their customer before you visit other stores-- so don't let the first place you visit get your name on the dotted line by tossing in extra goodies like a free power mike, or an antenna, etc.

Shop wisely, and you'll come up with your ideal rig.

9

Equipment Features

Let's take a random survey of some of the features found in AM and SSB transceivers:

40 Channels: All CB sets produced since the beginning of 1977 are required by the FCC to operate on all forty authorized channels. This, then, is a standard feature on all CB equipment when it leaves the factory. Sets with so-called "extra channels" or "HF capabilities" or the ability to operate on "upstairs channels" have all been "worked on" after they left the factory. Such equipment (supposedly) can operate on frequencies between 27.415 MHz and about 27.995 MHz. None of these frequencies are authorized for your use in North America, although some hobbyists seem to congregate there against FCC/DOC regulations.

Squelch: Because of the way sideband operates, a squelch may not prove to be of much value in that mode. But it's a handy aid for AM operations. You adjust the knob on your transceiver when no stations are being heard transmitting. At some point you suddenly hear the background noise vanish and the set seem to "go dead." Leave the control adjusted at that threshold point and you won't have to listen to the annoying background noise when nothing is happening on the channel. Any AM station that comes on the air who has a signal level stronger than the background noise will still be heard at normal volume. You may have to occasionally readjust the squelch due to changing noise level conditions. If you're a Sidebander, don't bother with the squelch-- because you'll miss out on half of what's taking place on the frequency you're monitoring.

RF Gain Control: This is a control that permits you to adjust the sensitivity of the receiver portion of the set. Unless a next door neighbor has a CB that's so close that he's knocking your socks off, it would normally be run a maximum gain. If run below maximum gain, you'll kill the set's ability to hear weak stations. Actually, this control isn't even included in most equipment designed for AM-only use. For Sidebanders, it does serve a purpose that takes the place of the squelch control. In SSB, the control can be adjusted to take out some of the background noise and reduce any adverse effects of having other operators in close proximity. You just back down on the RF gain to a point where incoming SSB signals can be copied with ease, and you'll eliminate a lot of unwanted sound. For SSB ops it's an asset.

Clarifier: It's a control that is found only on sideband rigs. This allows fine tuning of received transmissions so that the voice achieves its most natural sounding point. Normally, it receives $\pm$800 Hz from the transmitter frequency ("center slot"). All currently manufactured SSB rigs are "rockbound on center slot," which is to say that adjustment of the clarifier control doesn't change the transmitter frequency, only that of the receiver. A "slider" is a non-factory modification that "unlocks the clarifier" to permit the transmitter to change frequency along with the receiver. Although FCC regulations don't

CB Rigs They'll Never Design

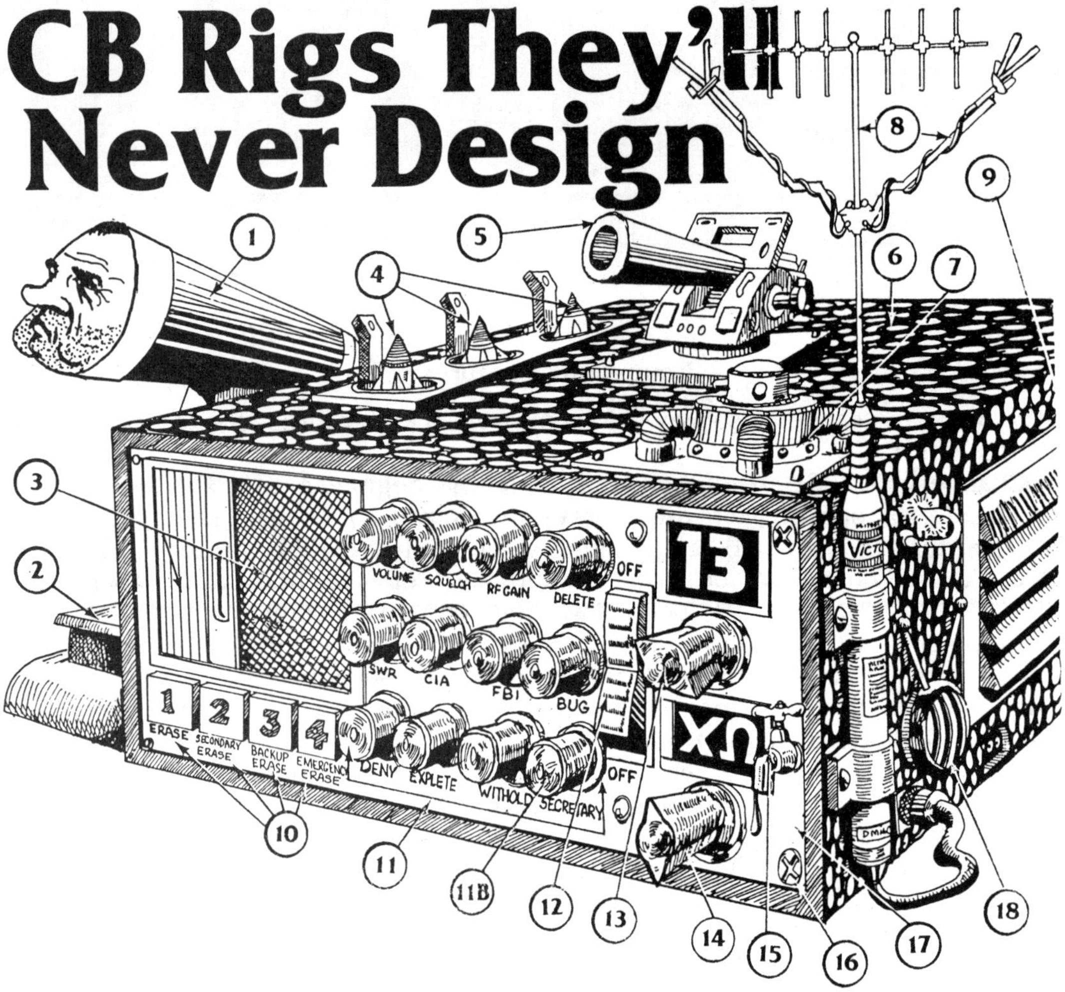

For the past few years the Uniden Company has produced some interesting and attractive CB gear in a series of transceivers wherein each set is named after one of the more notable Chief Executives of the United States-- Washington, Madison, and Grant, for instance. When I saw this, my devious brain immediately went into gear and came up with ideas for potential rigs for inclusion in this series that Uniden will probably never bring out. Like the **Coolidge**-- turn it on and it won't modulate; the **Ike**-- turn it on and it plays golf; the **Ford**-- turn it on and it falls off the table; the **Truman**-- turn it on and it cusses at you; the **Hoover**-- turn it on and it sells apples.

But the one I think would probably have the best sales potential is the **Tricky Dick.** This one has all of the best features built right in, as indicated in the artist's conception of how it should look. #1 is the patented Executive Privilege mike with the (#2) transmit bar that doesn't work until you push it hard three times. The speaker (#3) has a sliding door to muffle the voices. The devices shown as #4 are mini-missiles-- heat seeking jobs that search out the hot air coming from channel hogs and silence them once and for all.

The object at #5 is a small howitzer that protects the rig from theft and, even worse, unauthorized use. The case (#6), is 1-inch thick solid lead and can't be penetrated. The gizmo at #7 is the combo Promise Maker and fertilizer spreader. The built-in antenna (#8) is the exclusive "V for Victory" design inspired by the well-known hand gesture. At #9 we have a most unusual vent used for venting hostilities and frustrations. Most useful are the full array (#10) of various types of erase buttons "just in case."

To further aid your operating, at #11 are special controls to help adjust all comments until they are "perfectly clear"-- the top row controls normal CB'ing, middle row for abusive CB'ing, bottom row offsets the problems caused by the top two rows. A special "Secretary Button" (#11-B) causes 11 minutes of blank transmission when accidentally pushed. Power switch (#12) can only be moved from "off" to "off." Channel selector (#13) shown set to priority Channel 13. However #14 selects undercover channels reserved for insiders. Any information that needs to be leaked can be fed through the tap at #15 which the plumbers have installed. Special locking screws (#16) can be opened only by a crew of inept clowns carrying lockpicks. The front panel (#17) is reinforced steel. A hidden bug (#18) can be pressed into service to see if you are splashing over on other channels-- if you aren't, you can turn up the mike gain until you are.

The **Tricky Dick** comes with 500 QSL cards having your CB handle censored off all of them-- they will be sent to you in an envelope that has be pre-opened by unknown persons. You also get the logbook and matching pen with the disappearing ink. The rig is decorator designed in gleaming whitewash.

permit sliders (even when the transmitter slides within the authorized frequency tolerance), it has been a popular modification. Some SSB rigs don't lend themselves easily to being slid, and a botched slider modification is a disaster all around and can really screw up the set.

Delta Tune: About the same as a clarifier, except that the term seems to be used exclusively on AM-only equipment.

USB/LSB Selector: On SSB equipment (only), it selects the desired mode and even lets you switch over to AM-mode. The majority of sideband operations on 27 MHz are traditionally conducted in LSB. When a piece of CB equipment is described as having 120 channels-- and you know full well that there are really only 40 CB channels-- the description relates to 40 channels of AM capabilities, 40 channels of LSB capabilities, 40 channels of USB capabilities. It's just some harmless puffery that means little; AM'ers seldom operate on the frequencies where Sidebanders congregate, Sidebanders (who almost always use LSB exclusively in the band) generally stick to only a couple of frequencies, and the only people who repeatedly use USB are a few local nerds who take some small amount of delight in being perverse.

Noise Limiter /Noise Blanker: These two things aren't exactly identical, and some equipment offers both. In some units they are wired into the "on" position and can't be turned on/off by the operator. Although I prefer to be able to play with the little switches, so long as a noise blanker is designed into the rig it isn't absolutely vital that it is under the operator's personal control. The limiter is useful in silencing bursts of atmospheric static, such as from lightning or the type of static electricity noise that comes through on a hot, humid summer evening. The noise blanker seems to do its best job on impulse-type noise generated by automobile engines and electric motors. Use of the noise limiter may cause some audio distortion on incoming signals. Depending upon its location in the circuit of a transceiver, use of the noise blanker may reduce selectivity (ability to effectively reject strong adjacent channel signals) and/or make the equipment vulnerable to strong-signal overload. The ability to switch these functions in/out of the circuit lets the operator select his own approach.

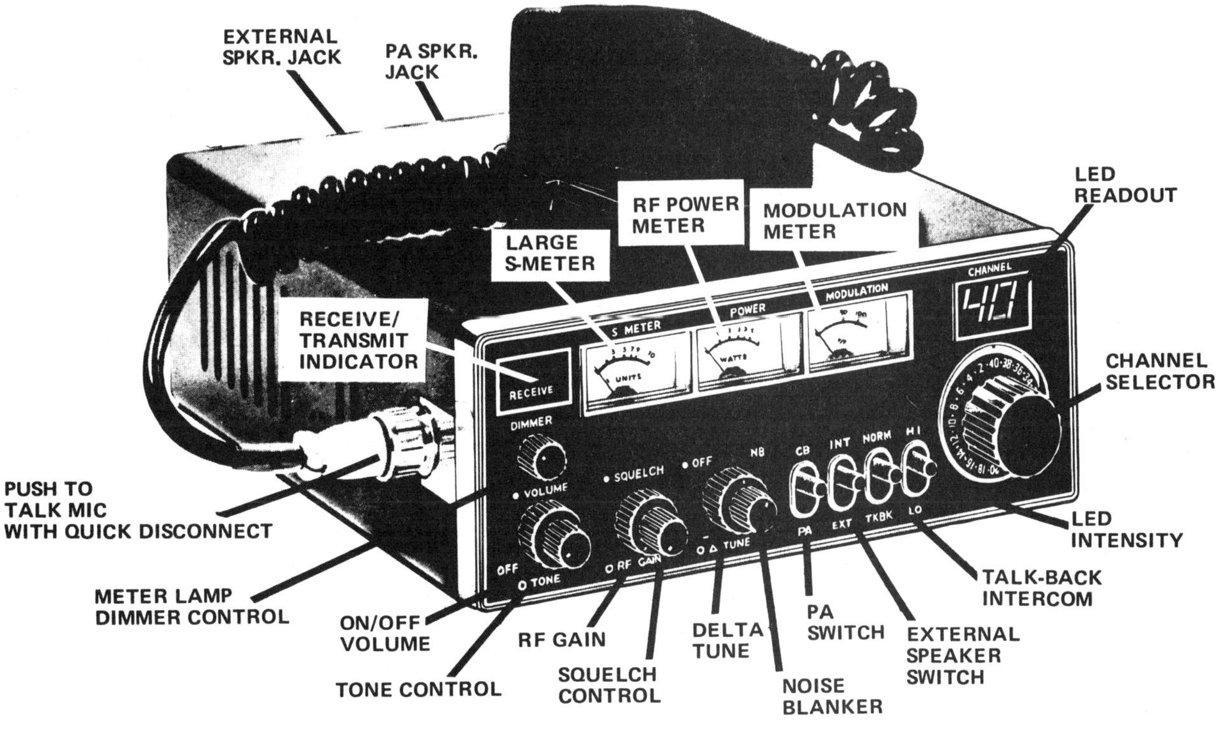

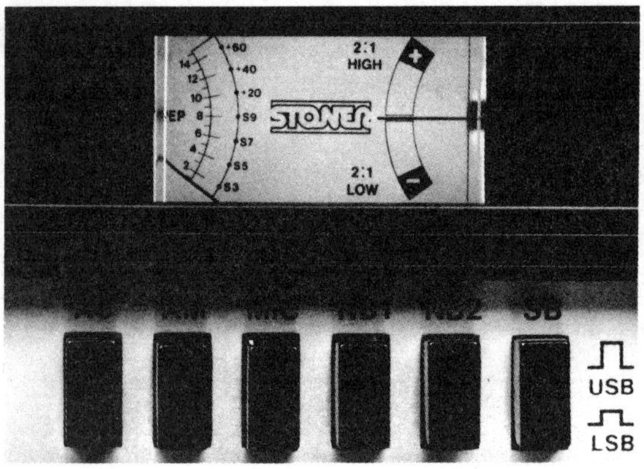

This closeup of a portion of the front panel of a Stoner PRO-40 deluxe SSB rig shows the unit's multi-purpose PEP wattmeter, S-meter, and advanced SWR bridge. The three buttons at the right are the noise reducers and the LSB/USB mode selector.

S-Meter or Indicator Lights: Such devices, which give a visual indication of the relative strength of incoming signals, will be of minimal value to Sideband communicators since SSB signals cause the indicators to bounce wildly up and down with each sentence. If you're trying to aim a rotatable beam antenna at another SSB'er, you may find it possible to direct the beam to the point where the deflections are at their highest point. For AM-side ops, at least you have a steady carrier to deal with, and that means the indicator doesn't bob around. That means you can give another station an S-meter reading on the signal you're receiving. Sidebanders don't give S-meter readings and may very often react to requests for same with some annoyance. The meter looks nice anyway, and visitors to the shack are always impressed.

RF-Meter or Indicator Lights: Similar to an S-Meter (and may be a combined function with the S-Meter), it lets you see a visual representation of the signal output from your transmitter. Relatively steady for AM transmissions, it bounces around wildly for SSB'ers. If you can't get any other station to answer your shouts, at least it lets you know that your transmitter hasn't gone totally dead on you-- perhaps it will enable you to have that information without the need to ask for a dreaded "radio check." Other than that, it's nice to look at but mostly useless, and it doesn't tell you how effective your entire communications system (transmitter/coaxial feed-line/antenna) is functioning.

Automatic Gain Control: Another AM-oriented device, intended to help weak incoming signals while also dampening the spirits of very strong local station signals that could overload the receiver. For sideband operation, it is best left in the "off" position.

PA Capability: You find this on many mobile units, probably because it costs the manufacturer no more to add it than leave it out. Unless you're installing the CB on your boat and need a loudhailer, or you are using the CB in a police or other emergency vehicle, it's of little use. Most states don't allow PA systems in private vehicles without a permit.

Tone Control: This is a knob that permits the operator to vary the bass/treble ratio of the received voices. Not many sets have this control anymore, but it is a nifty little extra. SSB'ers can use it to help knock out some of the background noise.

SWR Meter: Not something you'll use too often once your system is installed and operating, unless you want to check everything out after a wind or ice storm has had a go at your antenna system. If you don't have a built-in SWR meter, it's worth keeping one around the shack for such uses for periodic peeks at how your antenna system is stacking up. The higher the SWR, the lower the efficiency of your antenna system.

External Speaker Provision: I'm fussy about the sounds coming out of a CB rig and want them to be as readable as possible. Usually the speakers in smaller CB rigs, especially, the sound leaves something to be desired.

An external loudspeaker plugged into the rear panel jack provided always seems to improve the quality of the sound. Don't use a speaker intended for high-fidelity music reproduction, get one designed for communication or general purposes.

Ceramic or Crystal IF Filters: These are internal features that aid in improving the receiver's selectivity. Although not included in every transceiver, in SSB rigs it's an excellent feature.

CB Override: A feature on some of the fancier mobile transceivers; you get AM-mode CB radio along with your ability to play tape cassettes and also hear AM/FM broadcasts during periods when there aren't and transmissions taking place on the CB channel to which you're tuned. When you're hearing the radio or tape deck, the override feature gives priority to any CB transmissions-- which will come through and cut off the tape or broadcast radio reception temporarily. Personally, I'm not at all turned on with a CB set that cuts back and forth between CB chatter and music, I find it very weird and annoying. Unless you are monitoring CB in a rural area where only there are only a few CB transmissions a day on your favorite channel, you'll probably find it annoying, too. The type of CB equipment featuring this nonsense is probably intended for installation in fancy chauffer-driven stretch limos and similar elegant wagons.

Channel Display: Older CB rigs and current cheapie units have a channel display that consists of a light shining through some yellow plastic upon which the channel number has been printed. Better sets have LED's, but some sets with LCD's are also noted. I prefer the LED's to the LCD's because they're easier to read under poor light conditions.

Microphone Gain: A control on the front panel of some rigs (usually SSB). You can vary the amount of audio being put through from your mike through the transmitter. AM operators and most SSB'ers usually run this at maximum.

Antenna Warning Light: Provided on a few rigs we've seen. Let's you know when you've either forgotten to attach an antenna to your rig, or when the antenna system has suddenly suffered major trauma. Since attempting to

Rigs from CB's early days had only a few basic controls to deal with. This Eico 777 had a 6-channel transmit selector, variable frequency receiver, of/on/volume control, and a combo squelch and crystal spotter. The spotter let the CB op locate the unit's transmit channel on the tunable receiver. There was also an S-meter. These were the typical controls on all cheapie rigs.

transmit on an antenna-less CB rig usually causes major damage to the transmitter section of the rig, this is a handy gizmo. If your rig has one and you see it start to flash when you key up your mike, better stop transmitting pronto and correct the problem.

Positive/Negative Grounds: This mobile unit feature permits installation of your rig whether the vehicle has a positive (+) or negative (-) ground system. American (and most foreign made) vehicles are made with negative ground chassis. Unless you are planning on installing your rig in a rather old vehicle, this feature won't mean too much. Usually, most modern vehicle installations require only that the black power cord be connected to any chassis (ground) point, and the red lead go (directly or indirectly) to the positive ("hot") battery terminal.

VOX: A feature especially intended for the SSB operator and therefore a definite asset should you decide to bother with VOX. When you activate the VOX, you no longer have to press the mike button to transmit. All you do is start talking and your voice is transmitted. When you stop talking, the equipment instantly switches into receive mode. That means that if two or more stations using VOX are in communication, you can converse in a normal manner-- as soon as an operator pauses, someone else can interject a thought-- just the way you'd do in person or on the landline. It's a dandy feature found only on some of the more expensive units, although most operators don't seem to bother using the thing.

Frequency Counter: The Cobra 2000GTL and a few deluxe rigs have this as a built-in feature. LED's readout the exact frequency to which the clarifier is set-- to the closest 100 Hz. For persons having unlocked sliders or variable frequency oscillators that permit transmitting on any off-channel frequency in the band, it is if definite value in keeping track of the actual frequency being used. For those having stock transceivers, it looks pretty and lets you know how far other stations are from center slot, as tracked by your clarifier.

Other Miscellaneous Features: This includes digital clocks, fancy cabinets and trim, heaphone jacks, and whatever. Don't forget that it's what's inside the rig that makes the real difference in reaching and copying those distant stations out there in the boonies.

10

Accessories

Something that enhances the operation of your station that isn't included with the transceiver comes under the heading of an accessory. There are dozens of different things available in this category. Included here are ones that deserve serious consideration, although it would probably take several books to cover everything on the market.

Meters: While some transceivers come with various meters, others have none at all. Three handy meters around a CB station include an SWR meter, an RF watt meter, and a Field Strength Meter (FSM).

As mentioned elsewhere in this book, an SWR meter lets you check the efficiency of your antenna system-- notifying you in an instant how much of the signal your transmitter is attempting to be sent out is being refused by the antenna system. The refused signal, unable to get out into the airwaves, simply hangs around on the transmission line looking for mischief. It can cause some TVI, and if the built-up signal is strong enough it can back up (like your plumbing) and cause a howling feedback sound when you transmit, and it can even destroy the final amplified stage of your transmitter. Therefore, every new antenna system, or system that has been modified or damaged, or gone through the trauma of a storm of any kind, must be checked out to see its reading on an SWR meter (also called an SWR bridge). Some of the things that could cause problems that would produce a high SWR include damage to the antenna, antenna of the wrong length, poor impedance matching

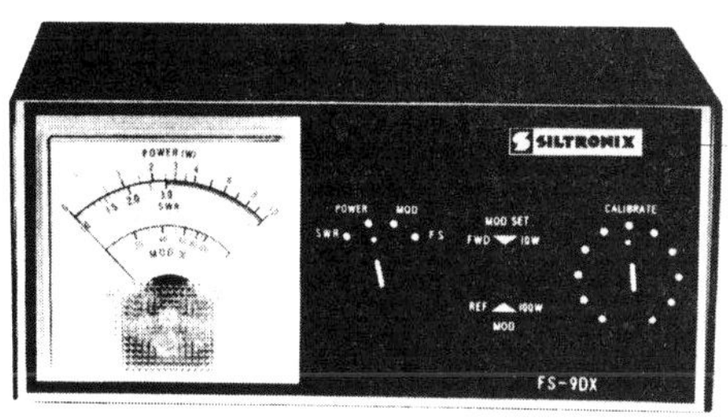

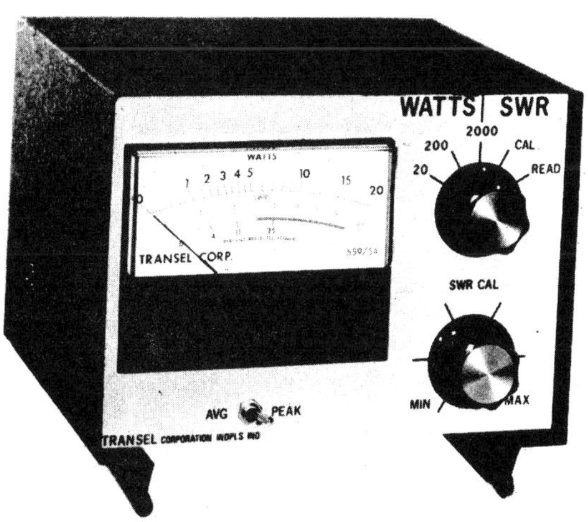

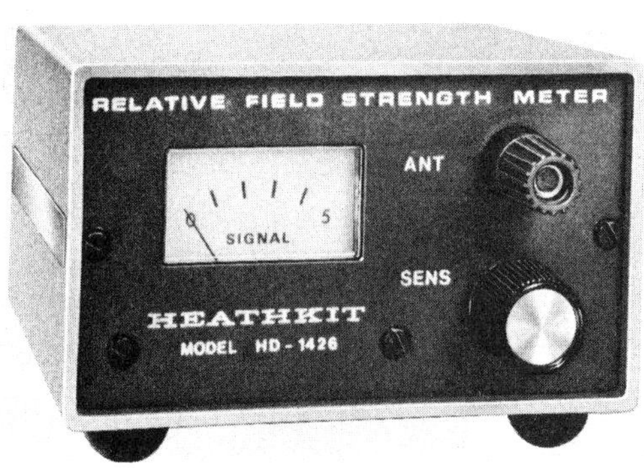

CB meters are available in single-function and multi-purpose types.

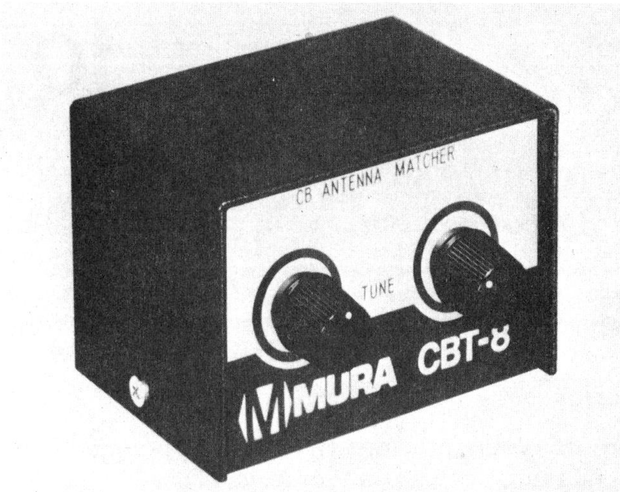

An antenna matcher or coupler permits you to reduce an SWR reading of 5:1 to a perfect 1:1 match, thereby protecting the transmitter from possible damage.

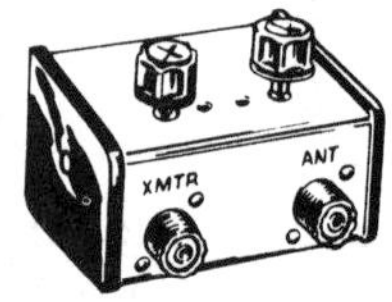

to the antenna, old or damaged coaxial cable, ice on the antenna, poor contact between the connector and the antenna, improperly installed connector, etc. Instructions for using an SWR meter come packed in with the device. It's simple. You don't have to be a fanatic about this, but it's good to check everything out once in a while. SSB stations will have to switch over to AM-mode on an AM-channel for an SWR checkout.

An RF wattmeter (sometimes combined with an SWR meter) lets you know your transmitter output power in watts. This is something that can only be checked with ease when you're transmitting in AM mode. If your meter says that your AM output is, say 2 watts, instead of the 4 watts you're allowed, then you're fighting the battle with only one fist. Time to have a technician see what the problem might be.

The FSM shows the relative strength of a signal emerging from a nearby transmitter and should be placed somewhere close to the antenna system. All it really shows whether your signal at a given point is stronger there than it was at some other time. Some ops think that an FSM is more useful than an SWR at tune-up time-- with antennas tuned for best reading on an FSM rather than anything an SWR meter is going to report. It may actually turn out that the highest attainable reading on the FSM-- indicating maximum signal leaving your antenna-- doesn't coincide with the lowest SWR reading. A friend of mine found that he got the best reading on his FSM when the SWR reading was 3:1-- far past the point where most operators like to see SWR readings. But he wanted the station set up for best signal, and that's how he operates!

Matching Networks: Some operators, noting that their SWR is too high, will purchase a little device called a matching network or tuner. By hooking this into the antenna cable it's possible to twiddle with one or two knobs and get an almost perfect reading on an SWR meter, showing a better match between the rig and antenna system and possibly avoiding all of the pitfalls of a high SWR.

These gadgets were devised years ago in pre-CB days and were originally intended for hams who wanted to load their transmitters into random lengths of wire, drain pipes, bedsprings, and other objects of questionable resonance. But CB antennas are ideally supposed to be generally resonant on 27 MHz and, if properly installed, really shouldn't produce some startlingly high SWR reading. Such a reading indicates a problem that may not be all that difficult to correct. If you do have an antenna system problem, even with a matching network, you're not putting out your maximum potential signal. Best bet is to use one of these devices only as a temporary measure until you can correct the probem.

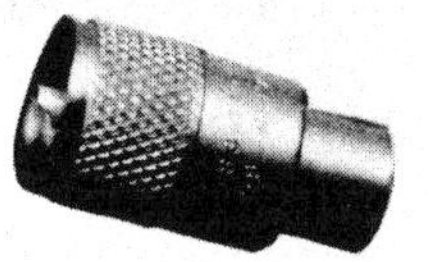

Dummy loads come in many versions. The one at the left handles only a few watts. At the right, this brute will sop up a full 1,000 watts and also provides a meter to confirm transmitter power.

Dummy Loads: A dummy load is akin to an ink blotter. It sops up and absorbs the signal emerging from a transmitter without sending it out into the airwaves. It fools the transmitter into thinking that there's antenna attached when there actually isn't one there at all. Avoids the disasters that can take place when attempting to transmit without anything hooked to the antenna terminals of the rig. The devices are primarily intended for those who are tuning up the circuits of their transmitters-- a task that should be tried only by a qualified communications technician.

Some SSB operators have discovered that dummy loads are also viable short-distance antennas, useful when communicating among several vehicles traveling together. Also they may be useful for communications between several base stations in the same neighborhood. They cut out incoming interference, keep the transmitted signal close in, and let you gab at length with your neighborhood Sidebanders without being accused of being a frequency or channel hog.

A wide variety of antenna switches are available that allow you to use your CB rig with any of 2 to about 6 different antennas. You can also take one antenna and switch it around from 2 to 6 different CB rigs.

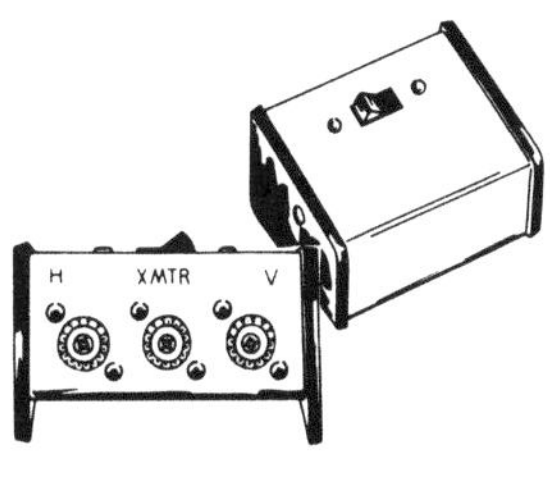

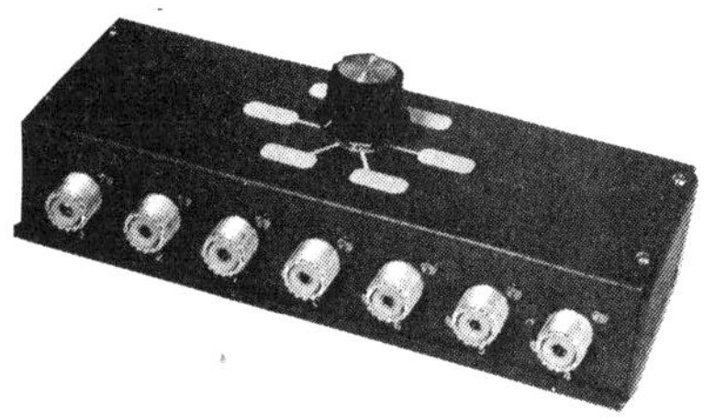

Switches: If you're running several rigs into one station antenna, or one rig into several antennas (such as a beam and also an omni-directional type), you're going to be faced with the prospect of continually unscrewing various connectors and then screwing everything back together again. Avoid this with multi-position switches designed for this purpose. There is some very slight detraction from your system efficiency, but so little that you won't notice the loss.

Coaxial Cable: Since it doesn't come with the set, I'd classify it as an accessory. For CB use, be certain you get 52-ohm type. Don't try using that leftover 72-ohm cable from your TV installation-- although it looks rather similar externally.

If you are faced with a run of 50 feet or less, you can stick with RG-58/U type cable. For longer runs, I suggest the heavier RG-8/U or some other

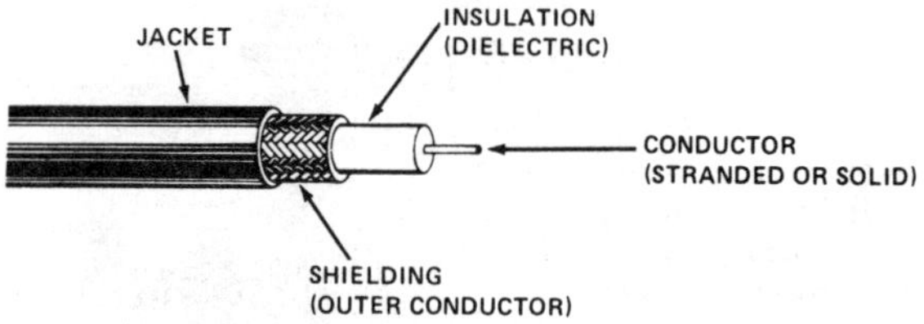

CABLE	RECOMMENDED USE	DIA. (IN.)	IMPEDANCE (OHMS)	CONDUCTOR TYPE
RG–58	mobile units	0.2	50	stranded
RG–8	base stations	0.4	50	solid
RG–59	co-phased antennas	0.24	75	stranded

At the left, the composition of common coaxial cable. At the right, some comparisons of several coaxial types. Note that co-phased mobile whips are one of the few times you'll ever need 75 ohm TV-type coaxial cable at a CB station.

low-loss type. When buying cable, get high quality stuff. Forget about the military surplus cable that's offered at bargain prices, just not worth the savings. Old cable, or cable that's dried out, and that has cracking, split, brittle, cut, or punctured outer covering should be avoided. Cable that you've been using at your base station for a few years should be replaced in the interest of better signals.

Filters: If you've been faced with television interference (TVI) problems, you know it's no fun. TVI from AM CB'ing causes more problems than from Sidebanding. While SSB signals have more kick than AM signals, the interference they cause to TV sets is usually too garbled for your neighbors to figure out what's being said. TVI from AM rigs pops in loud and clear on TV sets!

The suggested cure for TVI is to get a low-pass filter for your rig. The low-pass filters made for use with CB rigs filter out any signal products (such as harmonics) above 30 MHz. Since the TV channels start at 54 MHz, that's a real boon, since the second harmonic of a CB rig can put the whammy on TV Channel 2.

Low-pass filters connect easily to the antenna output of your CB rig and don't require any technical expertise or fancy tools. Theoretically, since all CB sets have been approved for sale by the FCC and are supposed to be TVI-free, there really shouldn't be a need for low-pass filters. Still, some transmitters are "dirty," especially those that have been had modifications to soup-up their power output or had extra frequencies installed-- and sets using linear amplifiers. It should be noted that low-pass filters are intended to handle only a specific amount of transmitter. So, if your pushing more

Even though CB rigs are designed to operate without generating TV interference, some do anyway-- especially if the CB's have been modified, are old, or if a linear amplifier is used.

power than a stock AM or SSB CB rig, then you'll have to get a low-pass filter. Popular models include the Barker & Williamson FL10/100 (handles 100 watts) and the FL10/1500 (will handle 1 kW).

Note that, although the instruction sheets don't say so, these filters work best when grounded to a nearby cold-water pipe with heavy gauge insulated bus wire.

At the TV receiver, a high-pass filter can also be hooked to the antenna terminals for an extra degree of blocking out errant signals that started out on CB. Ground the case of the high-pass filter to the chassis of the TV receiver. I'd say that you might want to buy one of these low-pass filters if you're coming through your own TV receiver, but don't spring for the cost of one for your neighbor-- you could find that you're then expected to treat all of your neighbors to the filters. Tell your neighbor that he's expected to see that his equipment is maintained in proper operating condition and that if it's picking up CB transmissions then he had better call in a TV service technician to fix his set.

Still, you (hopefully) aren't looking to go out of your way to annoy all of your neighbors by jamming their TV sets. If nothing you try seems to help, then you might consider staying off the air during the peak viewing hours-- roughly 7:00 to 11:30 PM. As for CB interference to stereos, telephones, electronic organs, hearing aids, music synthesizers, PA systems, and other such devices, filtering will have to be tackled by professionals.

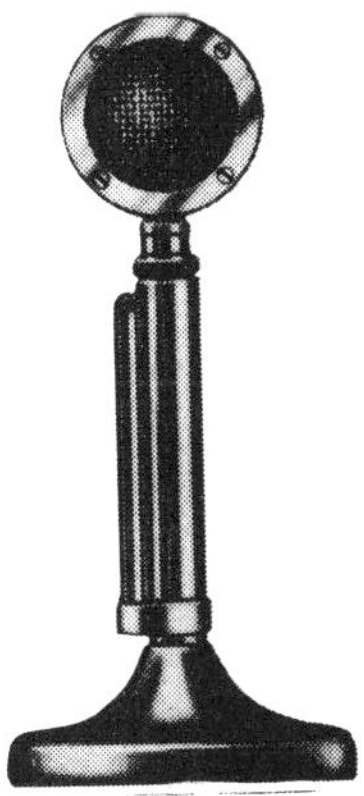

Two popular CB mikes. At the left, the famous Astatic D-104, now available in the original unamplified as well as an amplified version. At the right, the Astatic 1104-C, a really nifty looking (and performing) base station mike containing an adjustable voice booster (preamp).

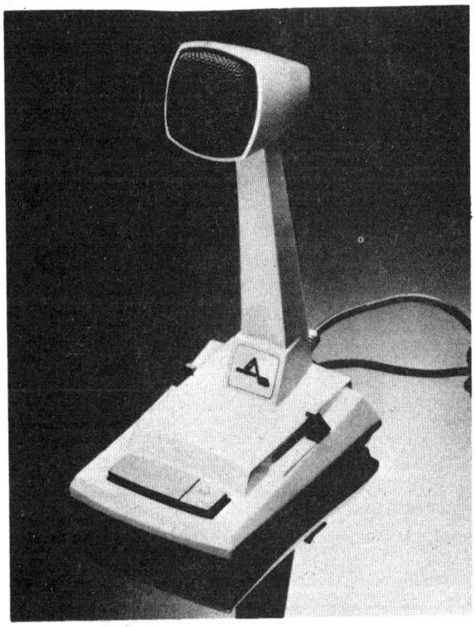

The Official 10-36 Machine

The Ultimate CB Accessory?

It may be that the ultimate cure for the annoying time check, commonly called the 10-36, has been discovered! Yes, all attempts at getting CB'ers to buy wrist watches have failed. Other action had to be taken.

The problem has been resolved with the Official 10-36 Machine, shown here. Here's what you do: Cut it out of this book and paste it to a piece of cardboard. Tack the cardboard to an easy observable spot near your operating position. Next time someone asks you for a 10-36-- pass along the reading from the machine-- exactly 10:36. You'll blow their little heads with your reply-- they'll think twice before asking you for anything as dumb as a 10-36 after 3 or 4 tries. When you've cleaned up your own local area, pass it along to a pal.

For base and (especially) mobile use, a combo headset/mike offers convenient hands-free operation. In a mobile installation, one of these gizmos is a definite safety asset, letting you keep both hands on the wheel. The transmit/receive button attaches to the steering wheel for activation by your thumb.

Microphones: Although a microphone comes with every CB rig, the replacement microphone industry thrives because CB'ers (especially Sidebanders) like to add their own personal touch to their station-- and this is a great start.

For one thing, not all rigs come with desk-stand type mikes. While a hand-held mike is handy in the mobile unit, it's not as convenient to use in the base station as one on a desk stand with a push-to-talk (PTT) bar. Also, some of these replacement mikes (desk and hand-held types) are available with extra amplification-- they call 'em "power mikes." These provide extra audio punch.

Popular non-amplified desk mikes include the Astatic D-104 and Shure 444-D. Power mikes that CB'ers especially like include the Astatic Silver Eagle, Astatic 1104-C, Telex-Turner Model Plus 3-B, and Shure 526-TS-2. Within the CB fraternity, several mikes are often considered to match up best with specific rigs to produce optimum results. This may or may not be true, but I do know that certain operators' voices sound better on one mike than another. Some mikes, like the old Astatic D-104 (lovingly known in CB as the "Lollipop" and the "Chicken Choker") and the Shure 444-D have achieved almost-legendary fame within the ranks of CB-land.

Some mobile CB'ers have opted for the use of a combo headset/mike since they leave both hands free to drive. If your SSB rig has VOX, then you've got it made-- you'll have a ball as you drive down the highway without having to grope for the mike, or getting the mike cord tangled in the steering

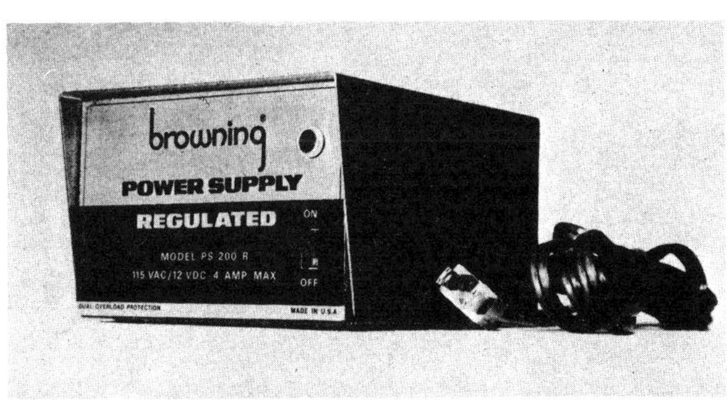

For running a mobile transceiver at your base station, you'll need a power supply, but make certain that it's a "regulated" power supply.

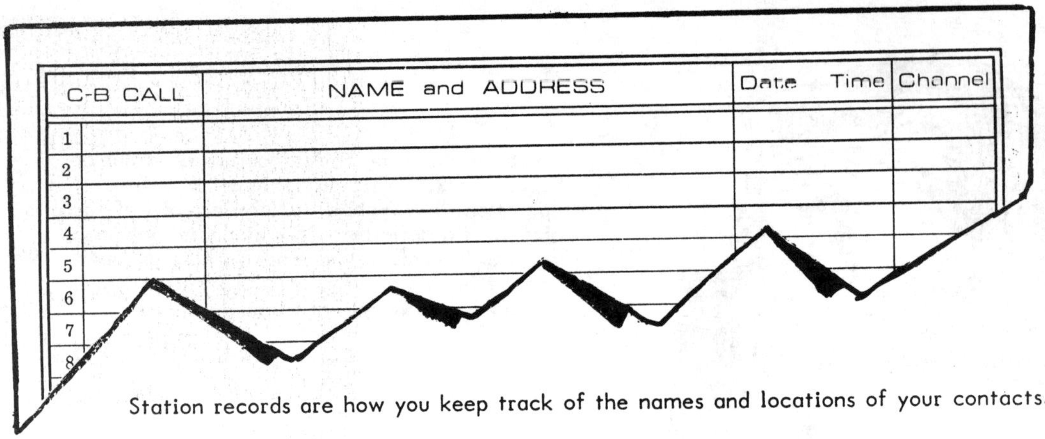

Station records are how you keep track of the names and locations of your contacts.

wheel. Remember the ode to the mobile operator:

He went barrelling down the highway,
One hand held mike and switch.
A truck pulled out in front of him,
He didn't even twitch.

Power Supplies: In order to operate a mobile transceiver as a base station from your home, vacation spot, or office, you'll have to get a power supply that will change the house current into the 12 VDC needed by the rig. You'll need one that can supply enough juice (6 Amperes should do it) and one that is regulated. Unregulated power supplies, while inexpensive, will cause your transmissions to suffer from a loud hum. Also, don't try to get away with using those little "battery eliminators" which seem to, at first glance, be suited to the challenge-- they aren't.

Station Records: You may find it handy to maintain a station log containing vital stats of other stations you contact-- handle or sideband ID number, his/her first personal (name), location, channel, his hobbies, occupation, and any other information that will jog your memory the next time you connect. Several commercially logbooks are available, or you can use the old reliable file-card system.

Mobile Mounts: Mobile CB rigs are always supplied with mobile mounts. For many people, these mounts are fine and dandy. But they've got their drawbacks. For one thing, they offer poor security-- the rig can be yanked

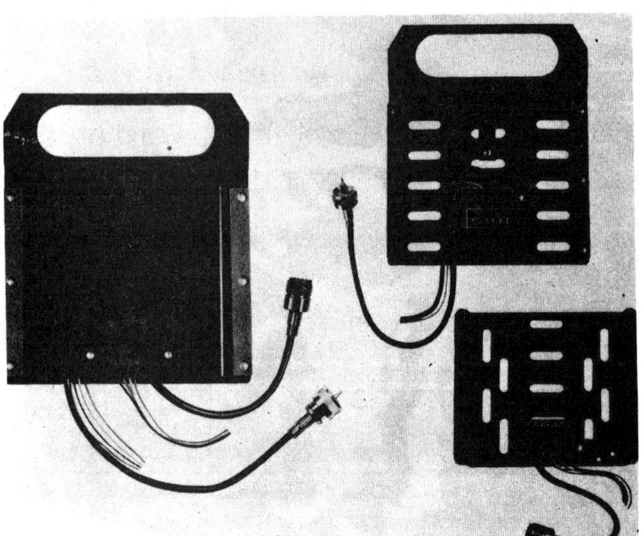

Mobile slide mounts replace the mounting brackets supplied with CB rigs, offering you the ability to easily remove the rig from 'neath the dashboard when you leave the vehicle.

One good turn deserves another, and if you've got a beam type antenna this becomes something to think about. Antenna rotators are rated for how much weight they'll be able to handle, so make certain that the one you get is up to the responsibilities of its job.

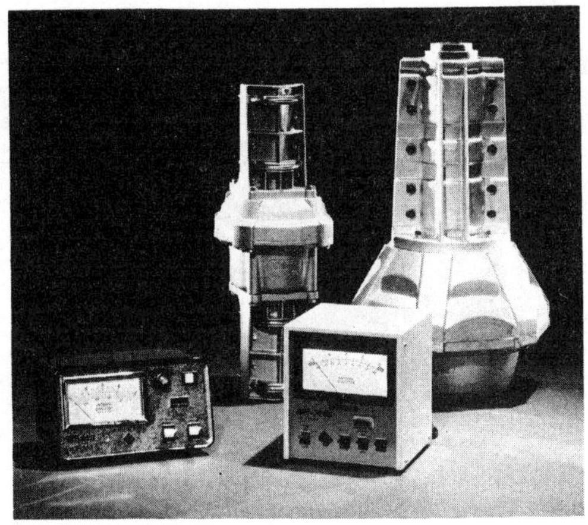

out of the vehicle by a thief without very much effort. That's because the only thing holding the rig to the mount is a screw at each side, and that's about the same thing holding the mount into the vehicle. Also, if you want to remove the rig from the car, it's a nuisance.

Many operators opt for a sturdy slide mount. This device attaches firmly and securely to the vehicle and has provisions for the CB power source and antenna connection. Another part of the slide mount connects to the CB rig itself, with attachments for the antenna and power hookup. Both parts of the slide mount easily fit together when the rig is installed in the vehicle for use-- better slide mounts have locks so that the rig can't be removed without a key. When leaving the car, it's easy to extract the rig from the vehicle by sliding it out of the portion of the mount affixed to the car. You don't have to fool around with any cables or power cords when inserting or removing the CB rig-- which can be hidden in the vehicle's trunk for security when nodody's around.

Phone Patches: A 'phone patch is a little gadget that hooks between your CB rig and the landline (telephone). With one installed, you can pass telephone calls to and from other CB stations, mobile and base. Someone at your base station is going to have to handle the duties of being the telephone operator (dialing numbers, etc.), key up the mike for each transmission, and that sort of thing. Still, it's a lot less costly than a cellular 'phone!

Antenna Rotator: If you decide to use a directional beam antenna, you're going to have to use a rotating device. This is a two-part affair, with the motorized part mounted at the top of the pole or mast upon which the antenna is mounted, the other part is located at your operating position and lets you select the heading you want the antenna to beam signals toward. If your beam is a 3-element job, you might be able to use a rotor intended for TV antenna use. On the other hand, those humungus multi-element super gain CB beams that toss out killerwatt signals are going to need something far more substantial than a rotor designed for TV antennas. Check the weight of the beam and see how it stacks up against the weight bearing capacity of the rotor you're thinking about buying. Also, take into account that if you live in an area where the beam will have to support wintertime snow and ice, it will substantially increase the weight of the beam. Note that, in very cold and icy weather, many antenna rotors freeze solid and refuse to budge a single, solitary inch.

Lightning Protector: Lightning protection is such a simple matter that it's amazing that every CB station with a roof mounted antenna doesn't have this safeguard. Several companies make the little attachments that go on your antenna cable and attach to a good earth ground. That's all it is, and it doesn't interfere with your signal. It's cheap-- a small fraction of what it would cost to fix the damage to your station in the event of a direct or nearby lightning strike.

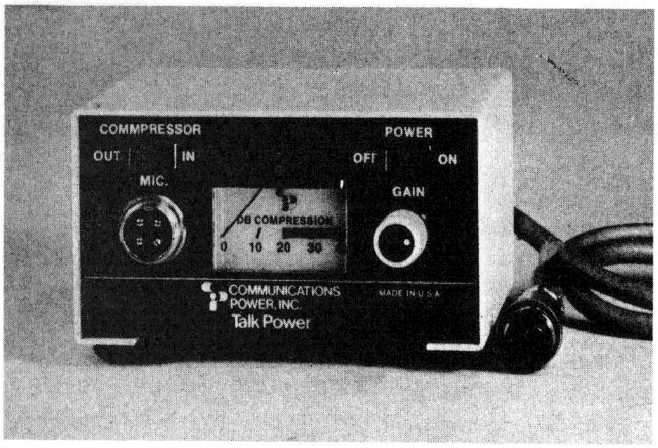

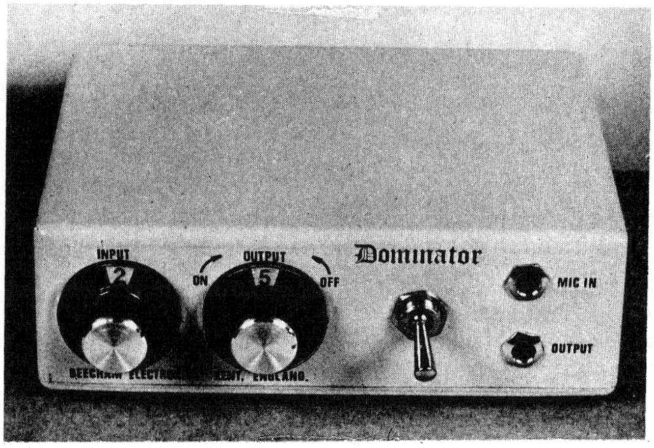

Processing devices of various types are available to reshape your voice for maximum talk power.

Speech Processor/Compressor: An accessory (sometimes included in the design of a power microphone) that improves ability of your voice to be more easily understood by distant stations and those copying you through interference. This may be done by increasing the average modulation percentage (changing the relative loudness of various voice components) and/or by reshaping the voice by filtering out certain bass (low frequency) voice products (this is called frequency compression). The amount of talk power is usually controllable by the operator, thanks to controls on the unit which prevent you from engaging in the horrors of overmodulation. By the way, some AM-mode stations seem to use echo chambers under the mistaken impression that they somehow enhance the station's talk power-- they don't. Other than making the operators sound like they're talking with their heads in a wastepaper basket, they make the voice extremely difficult to understand and annoying to monitor for more than a short period.

Headphones/Earphones: If your transceiver is made to be used with a headset, you may find these a good accessory when you're trying to read station that's giving you poor copy. They're also convenient when you want to operate in the wee hours without disturbing everybody else in your household. When the 'phones are plugged-in, the loudspeaker is muted. When buying headphones, do not purchase the type intended for stereo or high fidelity music reproduction. You need a set specifically designed for communications use since they offer a shaped frequency response intended to best enhance voice reproduction. Also check your transceiver's instruction manual to see the impedance value of the headphones suggested for use-- since headphones are available in 8, 12.5, 16, 600 ohm and other values. Excellent communications headphones ("cans") are made by Kenwood, Icom, Yaesu, Japan Radio, and several other manufacturers. Prices range from $25 to $70.

Selective Calling/Tone Squelch: A few transceivers are designed to be used, optionally, with devices that (when activated by the operators) won't permit the signals of random stations penetrate the squelch. The only stations that will "open" the squelch are those within your own network that are equipped with tone squelch devices. When a thus-equipped transceiver is keyed up, the transmission commences with a burst of coded tones that will permit the signal to get through the squelch circuits of other stations equipped with similar devices. If you've got a CB installation that you're using for business purposes, this would be a handy device since you won't have to hear any units on the air other than your own.

Receiver Booster/Preamplifier/Preselector: Although the designs may vary, the idea is to attach one of these little boxes in your antenna cable

In the late-1970's there were about 45 magazines hopeful of cashing in on the CB craze.

(usually where it attaches to the transceiver) and experience an increase in the level of incoming signal strength. Of course, it will also increase the level of any noise you're picking up through your antenna system. A good device, but make sure that you get one intended for transceiving and not intended for use with a shortwave receiver. Otherwise, when you attempt to transmit, the output of your transmitter will damage or destroy the device. A "bi-lateral" booster is another name for combination receiver/transmitter signal amplifier-- a device that isn't legal to use on CB since it raises the transmitter power above 4 watts. AM and/or 12 watts PEP on sideband.

Communications Receiver: Some fancy base station installations utilize a communications receiver to replace the CB transceiver's receiver portion. A communications receiver that offers a worthwhile improvement will cost somewhere between $700 and $1,200-- but if you're interested in pulling those weak stations out from under the noise and interference, it's the best way to go. Manufacturers such as Icom, Yaesu, Japan Radio, Kenwood, Ten-Tec and others offer a dazzling array of receivers from which to select. Since receivers tune the entire broadcast band and shortwave spectrum, you can also use the equipment for general communications and broadcast monitoring.

Magazines: At one time there were dozens of CB magazines available, although a couple were rather bizarre and seemed to exist for only a couple of issues. One magazine came out that was akin to one of those weekly supermarket tabloids-- with alleged CB scandals, attacks upon various companies and CB operators, and that sort of thing. By about 1979 virtually all of this had ended and the field had narrowed down to the two magazines that had been serving CB'ers since the 1960's, S9 Magazine, and CB Magazine. In early 1982, the owners of S9 Magazine transferred the publication to the owner of CB Magazine. Supposedly the intention was to bring out a combined publication, although neither magazine ever seemed to come out with another issue. As this book goes to press, 1 publication serves the US CBer.

Popular Communications is a monthly publication sold on newsstands, through many electronics shops, and by direct subscription. The magazine covers CB, scanners, shortwave listening, ham radio, and just about every

An add-on external loudspeaker can substantially improve the quality of the sounds you're hearing at your CB station.

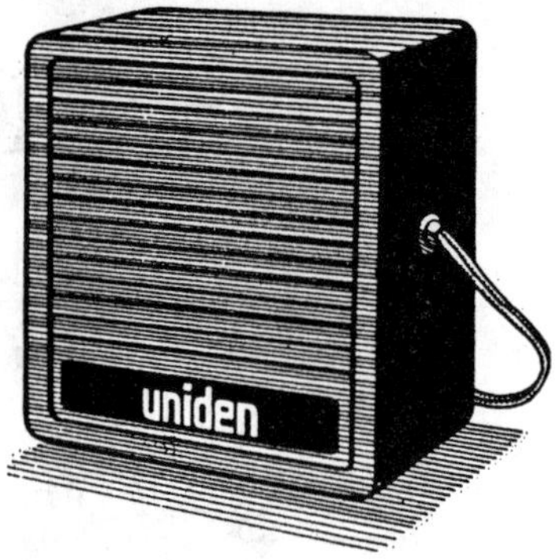

aspect of communications. The CB column is written by Bill Sanders, SSB-295, who wrote a column in S9 Magazine for many years. After putting in my own twenty years as Editor of S9 Magazine, I am the Editor of Popular Communications. The address of the magazine is 76 North Broadway, Hicksville, NY 11801.

External Loudspeaker: The loudspeaker that comes with most CB transceivers is small, but it is adequate for permitting comprehension of the incoming signals-- but some people want sound that's better than just adequate. If your rig has terminals on the rear panel for an external speaker, adding one takes only a few minutes work with a screwdriver and will substantially improve the sound quality of all incoming signals.

While it may hit you as a great idea, don't try to use one of the speakers from that old stereo that you put up on the shelf of the garage last year. Speakers intended for music reproduction don't do very well when pressed into service for two-way communications. Speakers intended for communications uses are designed for best audio response only in the frequency range of the human voice, thus eliminating various distracting sounds that could come through outside that range (such as would be reproduced by wide-range high fidelity speaker).

Some CB manufacturers offer external speakers that match (in external appearance) the company's transceivers. If you're not fussy about getting the looks to match up, then just get any small external mobile or outdoor speaker in a case available at any communications store, or ones made for use with ham transceivers.

11

AM/SSB Station Installations— Some Thoughts

Having operated 27 MHz rigs in just about every possible environment and under practically all conditions, I find that taking some extra thought and effort to assure ease of operation is of prime importance.

Even something as seemingly inconsequential as the location of a desk or table enters the picture. For instance, for comfortable operation, a desk or table should be about thirty inches high and have a large enough surface to accomodate your rig, a telephone, any needed books or reference papers, as well as an area for your microphone, your elbows, and and any papers you'll be using for notes or records. In fact, try to have at least sixteen inches between the front edge of the table and the control panel of the CB rig, with the rig placed slightly to the left of the desk.

If the desk is shallow, put the rig all the way over to one side and any books and papers to the left, leaving the center clear. It might be helpful to construct a raised platform for the rig, about five inches high at the front and three inches high at the rear. This will tilt it up for ease of operation and the area beneath the shelf can be used to store pencils and pads. Make the platform of half-inch plywood. Of course, you can always put the various

BASE STATION OF THE MONTH

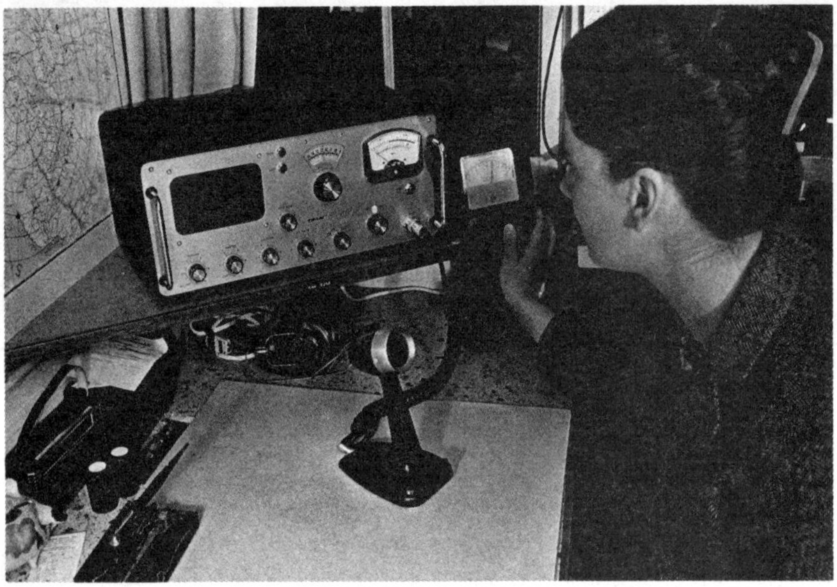

If there's a shortage of desk space, the transceiver can be located to the side on a shelf of its own.

components of your station on bookcase shelves if you don't have room for a desk, or if a desk would look out of plàce in the decor of your room.

Next, think about locating your station in a suitable and convenient place-- think about comfort, too. A damp and unheated basement is bad news. Take quiet (not only for you, but others in your family) into consideration-- that means the room for the family's TV watching may not be a good choice. The selected spot should have access to electrical power, also to a window (for antenna leads/light/ventilation).

As for safety precautions, bond all metal cabinets together with heavy wire and hook the other end to a good electrical ground such as a cold water pipe. Don't overload the electric circuits-- you may find it a good idea to have a qualified electrician run a single, well-insulated, heavy duty line to your operating position and fused directly from the house's circuit breaker box. This may be better than a maze of individual long wires running all over the room (via extension cords) to individual wall outlets.

I've always found it handy to keep a telephone and listing of useful telephone numbers (police, fire, ambulance, tow truck, etc.) off to the right side of the desk. Likewise, I've found it uncomfortable to locate the operating position too close to a radiator.

Other things you might wish to assemble at your base station include an easy-to-read station clock, a pencil sharpener, a large map of your area, and a desk lamp. Note that fluorescent lamps sometimes generate radio noise and you may be happier with a lamp having a 100 watt light bulb.

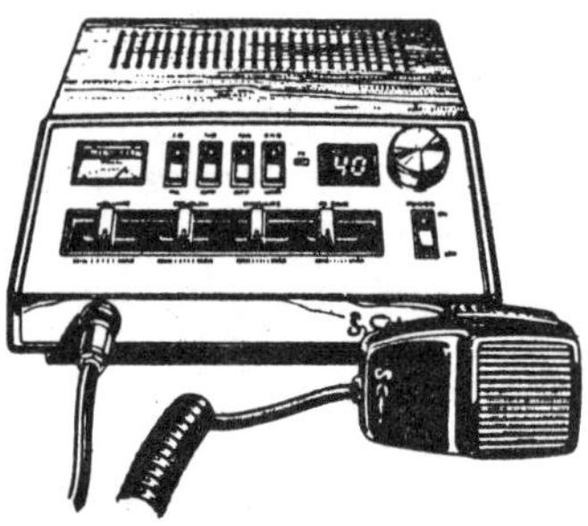

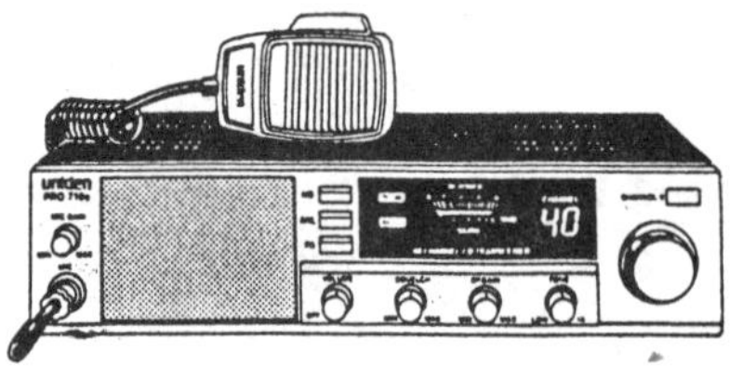

Three modern AM-only base station transceivers, the Cobra 90LTD, Uniden PRO-710e, and Midland 76-300.

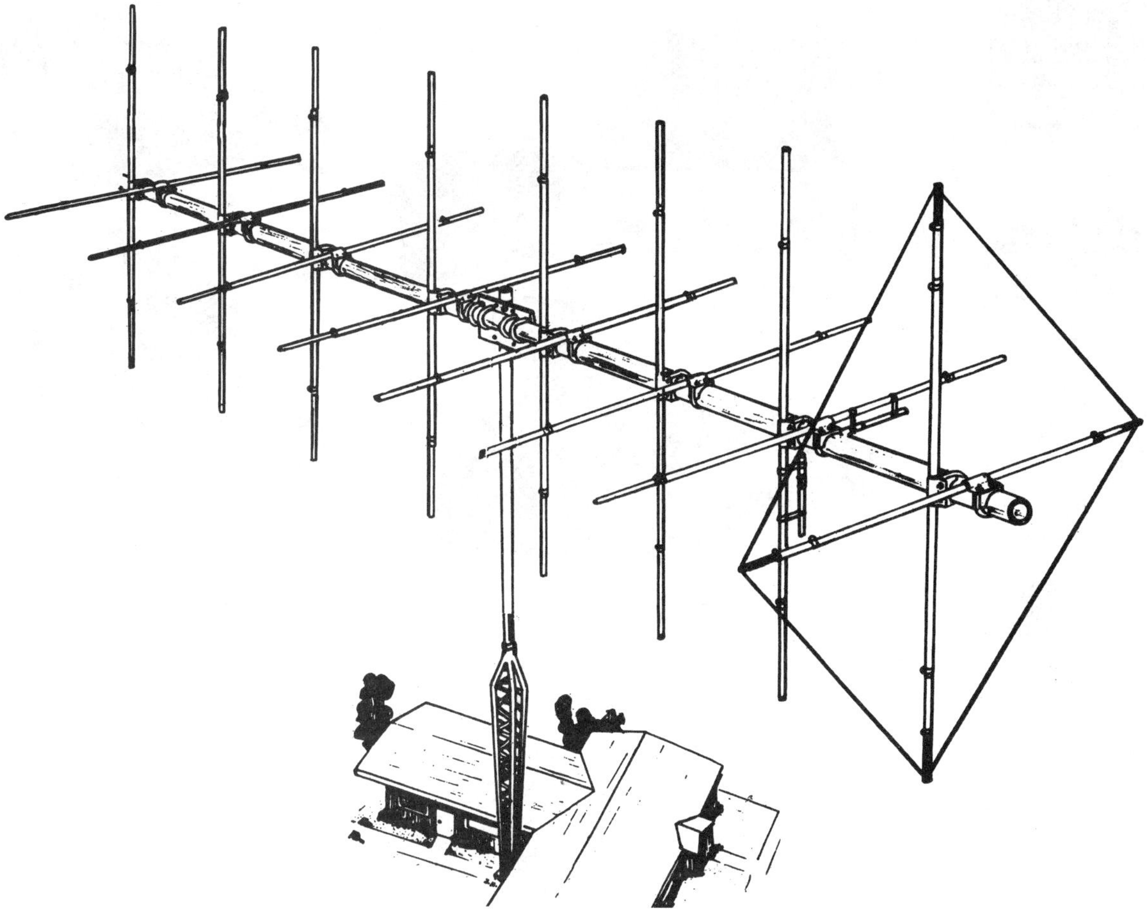

Antenna Installation Thoughts

Next, let's think about your base station's antenna installation. If you're going to utilize a beam in addition to an omnidirectional ("stick") type, you've got to prepare to different mounting locations. A chimney might be ideal for the omnidirectional antenna, but a larger beam type antenna should be on a well-guyed tower or tripod.

Check FCC and local zoning regulation relating to antenna heights.

Many operators go whole-hog with beautiful antenna towers as they do a good job and are very steady. You may be able to achieve similar results with with a smaller cash outlay if you use a telescoping-type TV mast. Whereas a 50-foot tower runs into several hundred dollars, a 50-foot telescoping TV mast is surprisingly cheap. Check them out at any TV shop, and they're also in the Radio Shack catalog. Fiberglass guy wires are preferred to metal since metal guy wires could distort the signal radiation pattern from your antenna. If you use metal, place glass or porcelan insulators in each guy wire at intervals of about 5 or 6 feet in order to reduce any harmful effects to your antenna's radiating pattern. All antenna installations should have adequate lightning protection.

An **important** word of caution: Do not mount or attempt to mount any antenna or tower where it could come into contact with electric power lines during or after its installation, or where it could topple onto the property of a neighbor. Each year there are numerous tragic accidents reported involving persons installing antennas-- persons falling from towers, towers falling into streets, roofs, on bystanders, and into electrical lines. Windy days are especially poor times to attempt to elect antennas. Be overly cautious in all respects when it comes to antenna installations-- you can't ever be too cautious!

uniden
Remote CB Transceiver

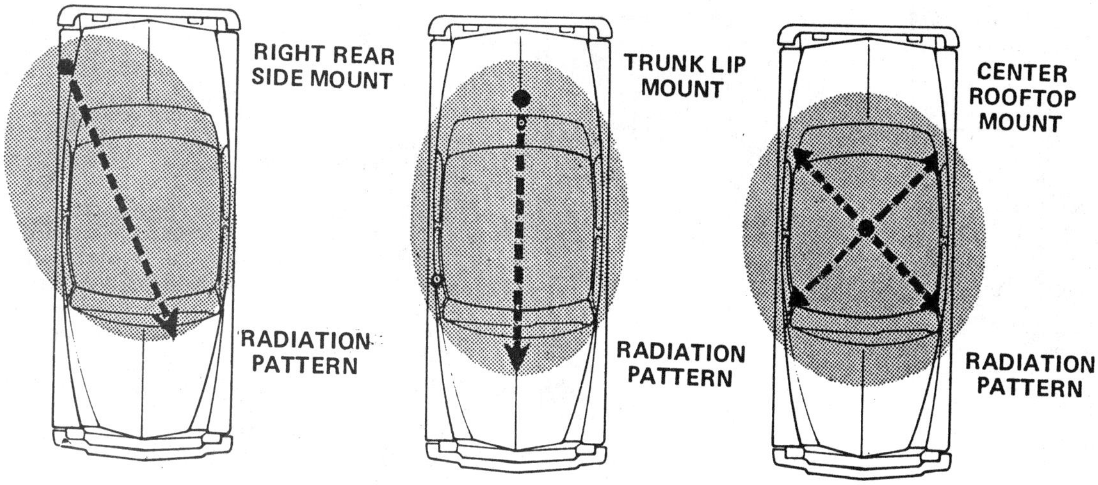

For maximum wind protection, here are some hints I learned while I was operating in breezy Oklahoma. If you have a beam, keep it aimed at the direction of the wind-- this lowers the surface the wind has to blow upon than if the beam is positioned broadside to the wind. If the rotor freezes up with ice or simply refuses to operate in extremely cold weather, you may want to consider not attempting to use the rotor until there's a thaw since repeated attempts to use it while frozen could cause damage. If your antenna is supported by a few sections of TV masting, use steel masting instead of the thin-wall aluminum type-- and you can add some strength to either type by dropping a broomstick down the center. All connections to the antenna should be covered with "Coax-Seal" or similar protective material (such as silicone paste) to seal out moisture and corrosion.

Mobile Installation Thoughts

I've always preferred to pull the power for my mobile units straight from the vehicle's battery rather than from the fuse block; for some reason it picks up less ignition noise. If you do get a lot of engine noise, the unit's noise blanker should help in reducing it, and also any noises sent your way from vehicles driving near yours. You can always, if the noise surpasses the ability of the noise blanker, install filters, coils, and traps on various noise generating parts of the vehicle such as the alternator and spark plugs. There are kits of parts with complete instructions available for reducing engine-produced radio noises. If you can't locate (or dont want/need) an entire kit, various individual mobile noise filters can probably be ontained from communications and automotive suppliers. Several such devices are always listed in the automotive catalog put out by J.C. Whitney & Co., 1917-19 Archer Ave., Chicago, IL 60680.

Mount your antenna as far away from the engine as possible, and (unless you buy a complete antenna/cable system), keep the lead-in cable as short pas possible. Don't buy a 25-foot length of cable and use it with the extra footage coiled up in the trunk or under the rear seat. While some pre-packaged

On the facing page is a galaxy of contemporary AM-only mobile transceivers. Column 1 (top to bottom): Cobra 19+, Cobra 18+, Cobra 21+, Cobra 25+, Cobra 40+, Cobra 29+, Cobra 20+, Midland 77-225. Column 2: Midland 77-104, Midland 77-106, Midland 77-149, Midland Powermax 77-250, Uniden PC-22, Uniden PC-43 marine CB. Column 3: Uniden PC-33X, Uniden PRO-540e, Uniden 520e, Uniden PRO-510e, Uniden PC-66, Cobra 67LTD.

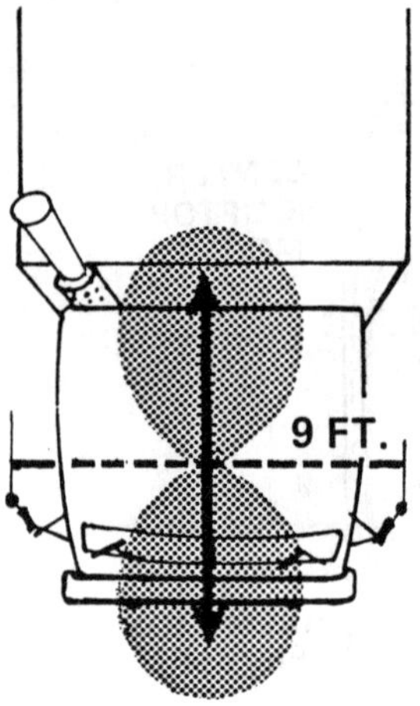

DUAL MOUNTS ON CAB OF TRUCK

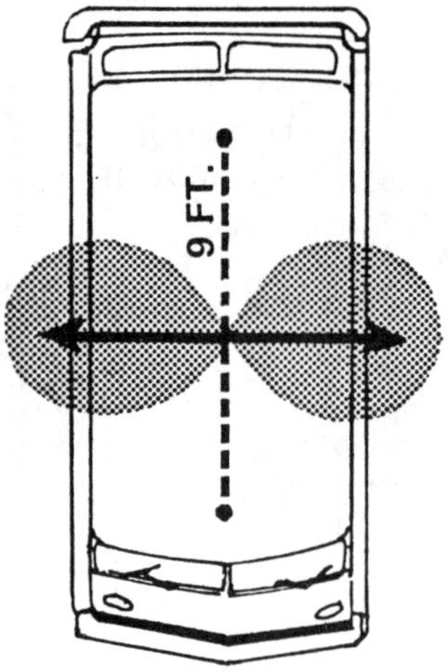

DUAL MOUNTS ON VAN

In order to have the expected forward/rear maximum signal coverage (shown at the left), co-phased twin mobile whips must be separated by exactly 9 feet (side by side). Only a large truck offers the opportunity to separate them this far. Placing them closer than 9 feet apart, or (as seen in the at the right) placing them 9 feet apart with one in back of the other, distorts the signal pattern to one that may be disadvantageous for best communications.

whip/cable antenna systems are designed with the need for the length of the coaxial cable to remain as delivered, generally speaking the cable doesn't have to cut to a specific length for maximum operation. A popular rumor!

Another unbased rumor is that a fiberglass whip doesn't perform at the same high level as a steel whip. Not true! They both work well. A full length (108 inch) whip, however, will outperform any type of shorter antenna made resonant by the use of a top, middle, or bottom located loading coil-- although most people use loaded whips because they are more convenient and don't bang into as many things like overhanging tree branches and garage doors.

When you mount your mobile antenna, keep in mind that the best transmission and reception will occur opposite where you've got the antenna mounted. Mounting it on the right rear will concentrate your best coverage towards left and forward-- for general Channel 19 use, when you want to contact oncoming vehicles for traffic information, this is probably the best spot.

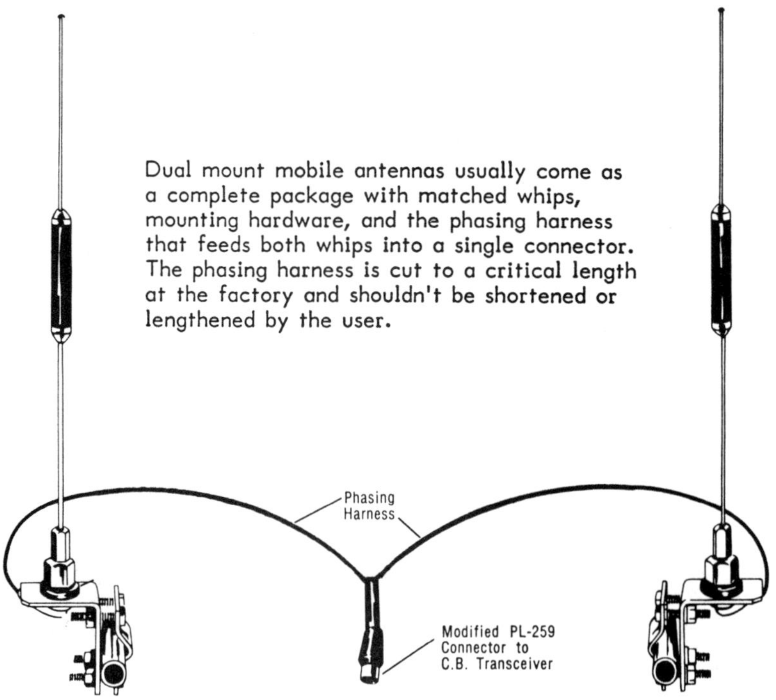

Dual mount mobile antennas usually come as a complete package with matched whips, mounting hardware, and the phasing harness that feeds both whips into a single connector. The phasing harness is cut to a critical length at the factory and shouldn't be shortened or lengthened by the user.

Omnidirectional (all-around) signal coverage is from the center of the roof, although it's not an especially convenient spot, and mounting it there (unless you've got a magnetic mount) is somewhat of a nuisance.

Magnetic mount antennas work just fine and stay attached to the vehicle no matter how fast you drive. There are also through-glass antennas available for CB, although I have generally been less than fully satisfied with their performance and ability to load up properly.

Twin-trucker type antennas are available for those who think that two antennas look better or work better than one antenna. These are fed from a "Y" shaped phasing harness that connects them both to your rig. They concentrate the signal towards the forward/rear of the vehicle and are popular with truckers driving large 18-wheel tractor/trailer semi-type trucks since they mount on the large side-view mirrors that stick out from the cab. This way, the wide trailer doesn't obstruct signal coverage towards the rear as would result with a single whip mounted on the roof or anywhere else on the cab. For passenger cars, use of this type of antenna arrangement offers no special signal coverage benefits but some operators (especially those driving vans and 4WD vehicles) like they way they look.

The mobile rig itself should be located in a spot where the controls may be seen and adjusted with a minimum of fumbling and distraction. The mike should be mounted where there is minimal danger of getting the cord twisted up in the steering wheel as you drive-- a most dangerous situation.

#1.

UNITED STATES OF AMERICA

OPERATOR'S CERTIFICATE OF SKILL IN RADIOCOMMUNICATION

This is to certify that, under the provisions of the Act of June 24, 1910,

____

has been examined in radiocommunication and has passed in

(a) The adjustment of apparatus, correction of faults, and change from one wave-length to another;

(b) Transmission and sound-reading at a speed of not less than fifteen words a minute American Morse, twelve words Continental, five letters counting as one word.

The candidate's practical knowledge of adjustment was tested on a "combination" set of apparatus.* His knowledge of other systems and of international radiotelegraphic regulations and American naval wireless regulations is shown below

Excellent.

(Signature of examining officer.)

Place Navy Yard, New York Date May 25, 1911

By direction of the Secretary of Commerce and Labor:

E T Chamberlain

Commissioner of Navigation, Washington, D C

I, ____________________, do solemnly swear that I will faithfully preserve the secrecy of all messages coming to my knowledge through my employment under this certificate, that this obligation is taken freely, without mental reservation or purpose of evasion, and that I will well and faithfully discharge the duties of the office

(Signature of holder.)

Date of birth __________ Place of birth __________

Sworn to and subscribed before me this 25th

day of May A. D. 1911

[SEAL]

Notary Public.

* It is not intended to limit the employment of the holder to a particular system, but merely to indicate the particular system in which he was tested for adjustment of apparatus

This certificate is valid for two years, subject to suspension or revocation by the Secretary of Commerce and Labor for cause. It should be kept where it can be shown to officers of the customs or other officers of the Government just before the ship leaves port

11—4308

This is a reproduction of the first radio operator's license issued by the U.S. Government, dated 25 May, 1911. This copy is missing the operator's name (at the top), and the signature/birthday/birthplace of the operator (bottom). Fill it in and you can prove to the world you were the world's first CB'er. Fill in the Patent certificate on page 34 and you also prove you invented CB itself!

12

Base Station Antenna Comments

The diligent operator realizes that the antenna system is one of the most things to consider when planning an effective communications installation. Selecting the best antenna system is an exciting challenge since it can make all the difference between average coverage and something that can lay a really significant signal to some distant point.

Gain

There are so many excellent books on antenna theory (written by experts whose highly specialized knowledge of the complex topic puts my own to shame) that I don't propose to embark upon a lengthy discussion of the subject-- but I would nevertheless like to point out when thinking about antennas, one of the names of the game is "gain." Gain is the antenna designers' way of denoting how much the antenna can multiply and boost signals. Actually, it's a measure of comparison, stated in terms of decibels (dB), which picks a reference antenna (a dipole) and says it has "unity gain." Unity gain means

Your CB shack doesn't need to look fancy, but unless you've taken best advantage of the possibilities for a decent antenna, your range will be less than great.

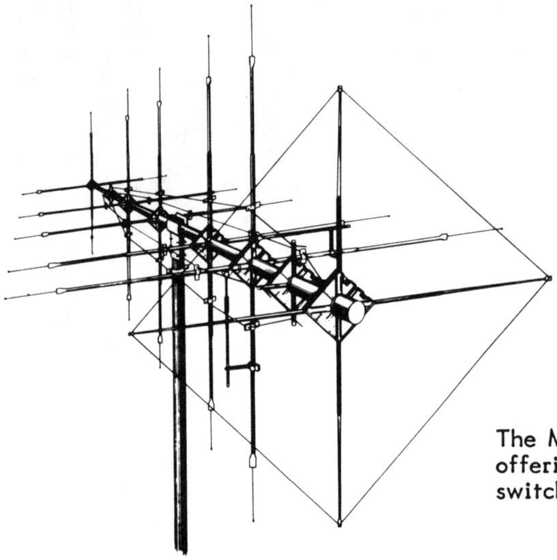
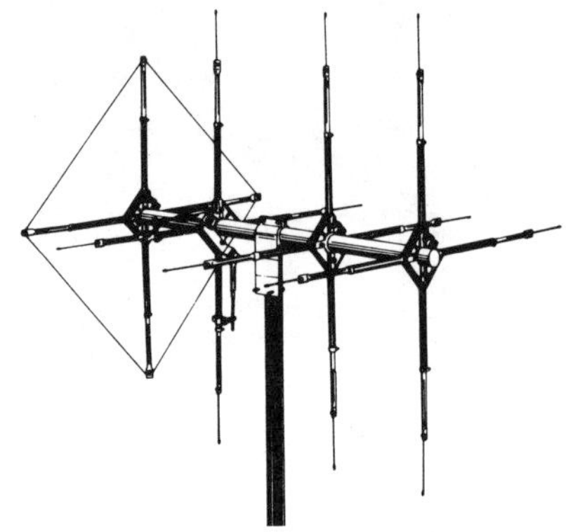

The Moonraker 6 (left) and Moonraker 4, two super-power CB beams offering 16.5 and 14.5 dB gain, respectively. Both antennas offer switchable vertical/horizontal signal polatization.

that the transmitter signal emerging from the antenna equals the one fed into the antenna; that is to say, the antenna doesn't offer any gain.

Antennas being tested for the amount of gain they offer are compared against a unity gain antenna, and the ability of the test antenna to beat the performance of the dipole is called the test antenna's gain. A 1-dB gain is another way of saying unity gain-- but a 3-dB gain means that the antenna took the signal and effectively doubled the transmitter power. So a CB rig that puts out a 4-watt AM signal will provide 8 watts of effective signal (legally) when fed into a 3-dB gain antenna. Sidebanders get an extra bonus since they are starting out with a signal that has 12 watts PEP-- a 3-dB gain antenna will parlay that into a signal that hits the band with 24-watts of clout. If that same signal can be run through an antenna with even higher gain, the potentials are startling. For instance, there are antennas that offer 14 dB of gain and claim to legally multiply a CB signal more than 25 times. A 12 watt PEP sideband signal will emerge as if it came from a 500 watt transmitter. Yes, I said it's all legal!

Of course, the more gain you have, the bigger, more cumbersome, and more expensive the antenna you'll be working with. Still, a tidy omnidirectional type antenna with some gain can be had from many sources at a reasonable price. One popular base station antenna claims to offer 3 dB gain and sells for less than $80 and it's no more awesome than a 16-foot pole you stick on your roof.

One popular arrangement at base stations is an omnidirectional (signal goes out about equally in all directions) antenna used in conjunction with a highly directional rotatable gain antenna. The idea is to monitor the channel using the omnidirectional antenna (sometimes referred to as a "stick") and then switch over to the directional beam antenna once a contact is estalished. A really good directional array is probably too directional for general use since stations calling you from areas outside its narrow coverage swath will probably not be heard at all. However, when you use it to transmit, it not only boosts and concentrates your souped-up signal towards the desired direction, it also attenuates (reduces) your signal in all of the undesired directions-- thereby reducing the amount of interference your station may cause in to other stations in those areas. In the case of Sidebanders, for instance, with its their signal clout-- ot "talk power," if you prefer-- it is something that should be seriously considered. If you're mainly interested in communicating with a particular station, let's say, to your northeast, why bother sending any of your signal over areas to the other points of the compass-- and why even bother hearing their transmissions? It's more considerate and permits

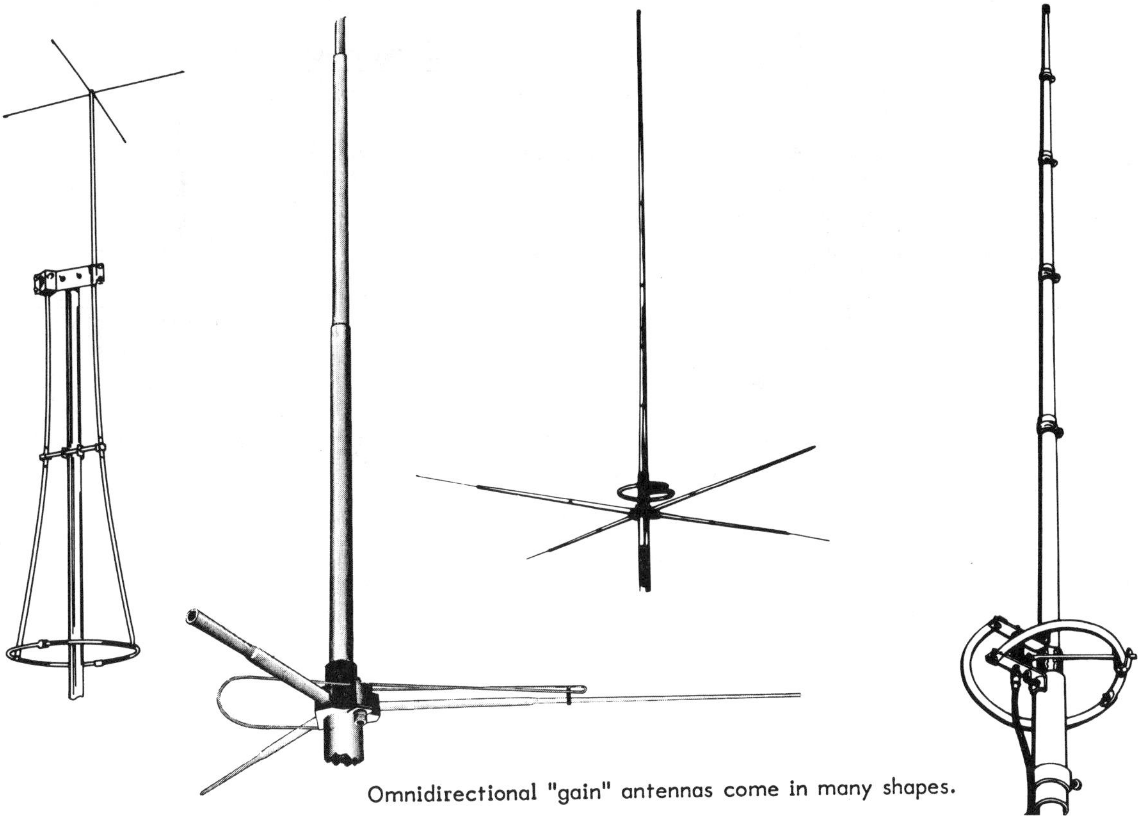

Omnidirectional "gain" antennas come in many shapes.

more area stations to use the same frequency simultaneously.

Omnidirectional antennas come in a wide assortment of shapes and sizes, including ground planes, coaxial sleeve, 1/2-wave, 5/8-wave and other types.

Directional beams antennas can get bulky, some are absolute monsters that claim a forward gain of up to 17 dB. Of course, beams can also be stacked one on top of the other to achieve additional gain over a single one of the same type. The huge monster beams require sturdy towers and powerful rotors. Their installation is often too ambitious a task for an inexperienced person or small group-- calling instead for the services of professional antenna installation firms who can deal with the many factors involved, including keeping it within FCC regulations and local zoning ordnances, as well maintaining the wide safety margain required. Of course, extreme caution must be exercised when installing or removing antennas. Contact with power lines by the antenna or any part of the mounting assembly can be lethal.

Antenna Placement

When it comes to signal coverage, base or mobile, height is the name of the game. Groundwave coverage of CB signals is basically limited to the horizon-- or "radio horizon" as it is called. That's the distance between the antenna and the horizon. The higher your antenna, the better your coverage, all other variables being equal. This isn't a major factor in skip communications, but it's a good rule-of-thumb when it comes to the signal coverage you'll have with mobile units and base stations in your local area.

If an antenna located ten feet above ground level is going to provide you with "X" miles of signal coverage, then the same antenna elevated to twenty feet above ground will provide more than "X" miles of coverage. In order to put a cap on the groundwave coverage potentials of stations in various radio services, the FCC establishes antenna height limits-- otherwise stations might presumably erect towers several hundred feet high and have the ability to communicate further than the FCC would like.

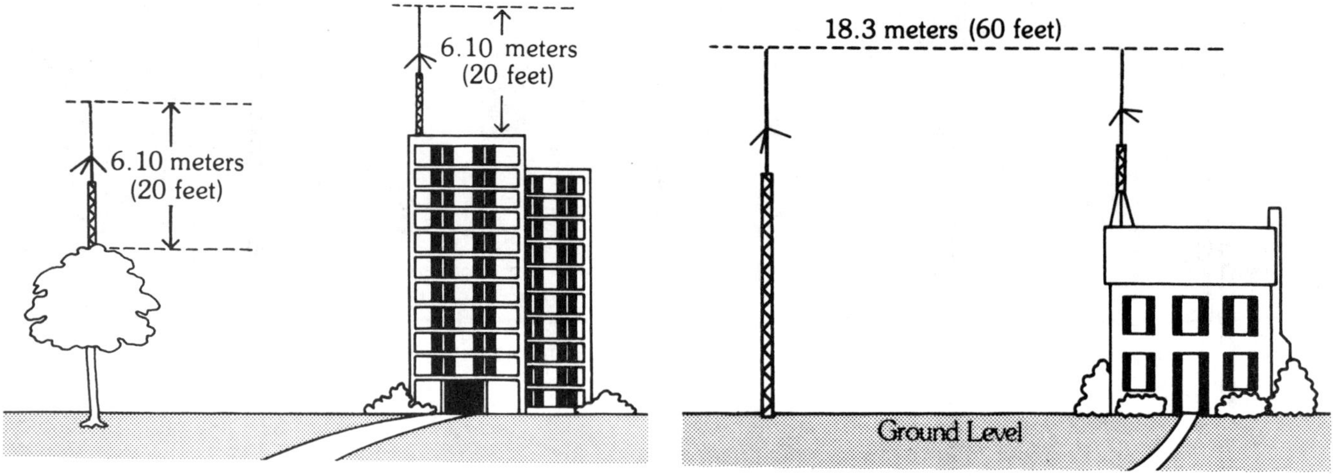

There are also local zoning and Federal Aviation Administration (FAA) regulations that can bear upon antenna heights. Of course, antennas can get some (legal) extra height if the stations with which they are associated are located at high sites-- like mountains, or in tall buildings. Not long ago I drove up to the top of Mount Equinox, Vermont (3,816 ft.) and met several CB'ers who regularly visit the spot because, even with a mobile unit, the signal coverage is fantastic! It's like having an antenna atop a 3,816 ft. tower (if there ever were such a thing). That's why all of New York City's TV stations all broadcast from antennas located atop the World Trade Center (1,350 ft. in height).

That's why CB'ers attempt to locate their base station antennas on the highest available mounting spot on their residences. According to the FCC's CB antenna height regulations (95.408), if you had your CB station located in a building that was 70 feet tall, the top of the antenna would be allowed to be 20 feet above the roof, giving you a legal-height antenna that was 90 feet above the ground.

If you lived in a two story home (about 25 feet in height), the highest allowable CB antenna couldn't extend to more than 45 feet above the ground if it were mounted on the highest point of the structure. Or, you could put a tower or mast in the yard and mount the antenna so that its highest point didn't extend past 60 feet above the ground-- of course, if your yard just happened on high ground-- like atop a hill or mountain-- you could add that height to figuring out how high your antenna was above seal level. A 60-foot high CB antenna in the back yard of a house in Denver has to be at least 5,340 feet above sea level, because Denver itself is 5,280 ft. above sea level.

Of course, we can't get too dazzled by such heady numbers. Your 5,340 ft. high Denver antenna system probably wouldn't give you any more range in working area mobile units than the CB'er with a 60 foot antenna located at sea level on at the ocean shore. That's because what's significant is the height of the antenna above the average terrain (HAAT). The 1,350 foot high World Trade Center looms 1,350 feet above the average surrounding terrain, so too the 3,816 altitude of the parking lot atop Mount Equinox-- and that lets the antenna send its signal to an extended radio horizon. In Denver, the average terrain all has a 5,280 foot altitude (that's why it's called "the mile-high city"), so there's really no benefit. A 60-foot high antenna rising above a backyard in Denver, is still only 60 feet above the average surrounding terrain.

Compliance with the FCC's CB antenna height limitations has never been very enthusiastic. There are many violations, and the FCC has hassled those caught with antennas exceeding the allowable height. In most of those instances, the violation is noticed only when FCC enforcement personnel

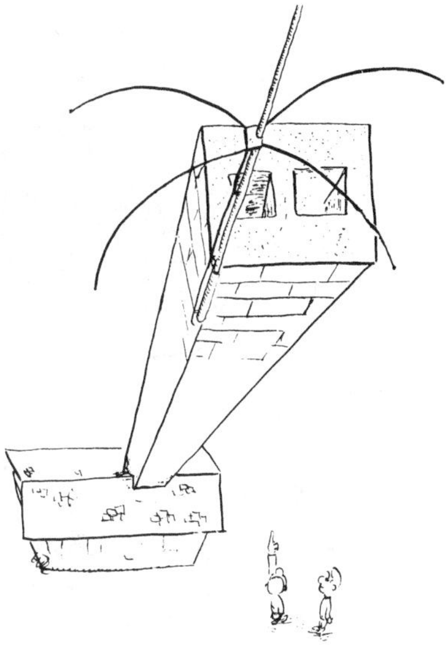

So what's illegal about my antenna?

visits the station to investigate other rule violations such as overpower or out-of-band activities, or after receiving interference complaints.

Computing Probable Groundwave Range

A person designing a base installation can calculate the probable range of their station. First they should find out whether the optical range includes the entire area to be covered. This is usually the case in a small city or town, and then it's easy to provide reliable 100% coverage.

The cheapest way to increase range is to elevate the antennas of those stations involved in communications. If, for example, an area lacks 100 feet elevation to clear the horizon, the elevation can be provided at the transmitting or at the receiving point, or can be divided between the two in accordance with the following formula: The optical horizon is 1.225 times the square root of the height of the observer plus 1.225 the square root of the height of the observed.

The radio horizon is computed the same way, but instead of 1.225 for the curvature of the earth, evaluate factors which increase the range and use a figure not less than 1.4 and, as a rule, not more than 2.0. It can, however, exceed 2.0 under exceptional area and moisture conditions.

Most computations are made from sea level or from the base level for a countryside. In practice there is always more height than that. Only a submarine barely emerging its periscope a fraction of an inch would be at true sea level. Even the height of a vehicle or individual, the height of the antenna, small or simple as it may be, atop the vehicle, the natural elevation of the ground at that point, all increase the horizon several miles. The first few feet of height are most important and provide much more range in proportion than twice that height would provide.

For example, here is a comparison of the computations to obtain the optical horizon and the minimum radio horizon of a mobile unit parked near an ocean beach-- that is, close to sea level-- and its base station in town. The figures used are merely illustrative.

Heights for mobile unit	Feet
Height of road above actual sea level	1.5
Height of car antenna above road	7.5
Total height	9

Formula: Optical horizon = 1.225 X square root of 9 = 1.225 X 3 = 3.675 miles.

OK, Charlie-- I just switched over to flat side!

Heights for base station	Feet
Height of land under building	55
Height of building above land	25
Height of antenna/mast	20
Total height	100

Formula: Optical horizon = 1.225 X square root of 100 = 1.225 X 10 = 12.25 mi.

Total optical horizon between top of car antenna and top of base station antenna = 3.675 + 12.25 = 15.925 miles.

To obtain the minimum radio horizon, use the constant 1.4 instead of 1.225. This gives the car a range of 4.2 miles and the base 14 miles. Total 18.2 miles instead of 15.925 miles. With good areas of conductivity, a constant of 2.0 is attained, which makes the car range 6 miles and the base 20 miles. Total range 26 miles.

Using the formula given, you can see the advantages of antenna height when calculating an aircraft flying at 6,400 feet altitude versus a ship having an antenna 64 feet above water level. For the plane, the optical horizon is 98 miles-- the minimum radio horizon is 112 miles. For the ship, the optical horizon is 9.8 miles-- the minimum radio horizon is 11.2 miles.

Even with low-powered equipment, the radio range would certainly be much greater than that shown, because of the ideal clear line-of-sight (LOS) path. Note the importance of the first few feet of height. It is logarithmic rather than linear in nature. The height of 64 feet gives a horizon of 11.2 miles. Yet 100 times that (6,400 feet for the aircraft) increases the range only 10 times.

Flat Side

"Flat Side" is a term sometimes encountered on 27 MHz, although more often on sideband frequencies than anywhere else. It refers to a signal that is horizontally polarized instead of the more popular vertical type of signal polarization used on 27 MHz.

CB antennas at all stations in a contact should be parallel to one another for best results. If one antenna in a contact is vertical (such as a mobile

In the 1960's, several flat side CB clubs were very active. AM groups used Channel 19 for this many years before the truckers took it over. Sideband base stations were experimenting with flat side for base-to-base operations, too-- but they were on Channel 16. This 1960's QSL is from a member of the Midwest CB'ers Channel 19 Horizontal Club.

whip), then all other stations communicating with that mobile unit should have similarly polarized (up/down) antennas. Since CB was originally devised as a mobile/base radio service, vertical polarization became the norm. A base station with a horizontally polarized antenna, lying over on its side (like a TV antenna) will find itself at a disadvantage in communicating with mobile units since they always have vertically polarized antennas. Flat side antennas are also directional (while vertical whips are omnidirectional). Even presuming that a mobile unit was driving within the signal pattern of a flat side base station, the disparity in polarizations of the two stations within groundwave range would cost as much as an S-unit in both directions of the contact (transmitting and receiving). (Cross polarization will probably be less of a problem with skip contacts-- it may even help in some specific instances; that's why some beam antennas offer switchable polarization to permit the operator to select the optimum polarization.)

CB'ers have long been tantalized by flat side and it's been an honorable form of experimentation dating back to the 1960's when several flat side clubs were formed. One, in the midwest, operated on Channel 19 many years before it became a mobile enroute channel.

CB'ers found that on long-haul groundwave contacts, base stations at each end of a contact can use flat side to substantially improve communications. For one thing, it reduces the stength of all of the other signals on the channel by about one S-unit, thereby bestowing what amounts to an effective one S-unit (6 db) boost to both stations' signals. As a side benefit, since much of the static and car ignition noise encountered on CB is vertically polarized, switching over to horizontal polarization stifles this, too! Of course, flat side is really practical only for base/base contacts.

Experimenting Is a Ball

For a small investment you can experiment with flat side, putting together easy-to-build horizontal beams. In addition, many operators are located in areas where both horizontal and vertical polarization are in regular use.

Flat side is especially useful for operators who live in sparsely populated areas. These operators come to learn that in order to reach a core of other AM or SSB operators, they've got to reach some regional population center off in the distance, but in one general direction from the operator. Setting up one of these antennas and orienting it in that desired direction will provide healthy coverage at relatively low cost-- so long as there are other stations off yonder able to switch over to horizontal polarizaton.

The Signal Sender

You can probably build this antenna in any of several versions for less than $15. It offers gain and directivity.

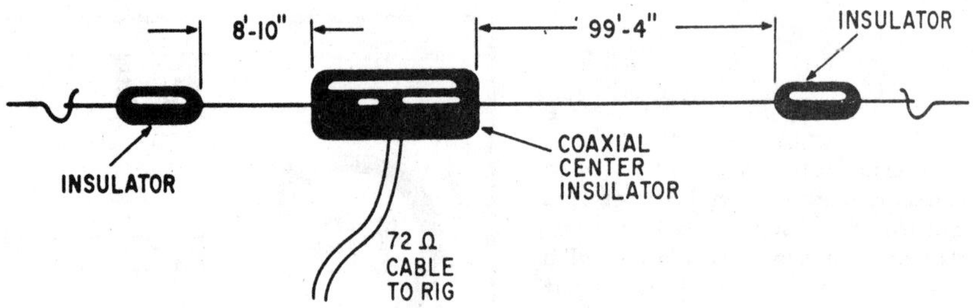

The Signal Sender: typical construction.

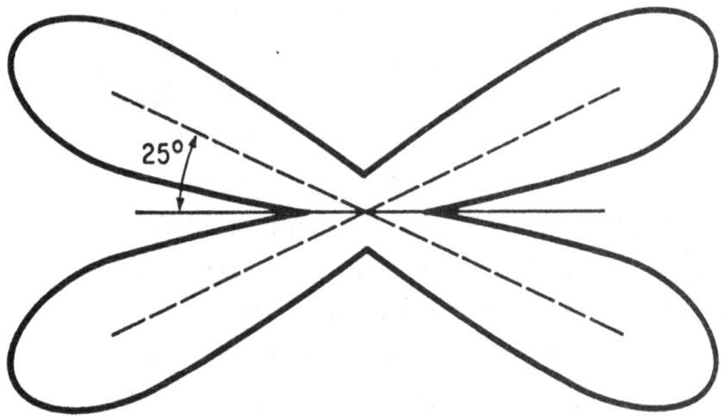

The Signal Sender: typical signal pattern.

The smallest version of this antenna is 3 wavelengths long and is fed by 72-ohm (TV type) coaxial cable at a point 1/4-wavelength in from one end. The exact measurements are shown in the diagram (but cut your wire a few inches longer to allow hooking to the insulators-- the measurements shown refer to what's required after the hookup). This version of the antenna produces a signal that is roughly shaped like a four-leaf-clover, with lobes extending 25° off the sides of the wire. If, for instance, you wanted to drop your signal into an area that is 52° magnetic from your station, then you'll have to orient the antenna either 47° or 77° to position the signal where you want it. Since the antenna can't be rotated, once you pick the coverage areas you're stuck with them.

When you've selected the physical direction for the antenna, place a 20-foot mast in the ground at least 109 feet away from your operating position. This mast is the "far" end of the antenna position (an old wooden rung ladder makes a good mast). Bury the bottom 3 feet of the mast solidly in the ground to hold it erect, guying it if necessary. A tree located between 109 and 120 feet from your operating position can be used instead of the mast-- just make sure that the antenna mounting point isn't more than 20 feet above the ground.

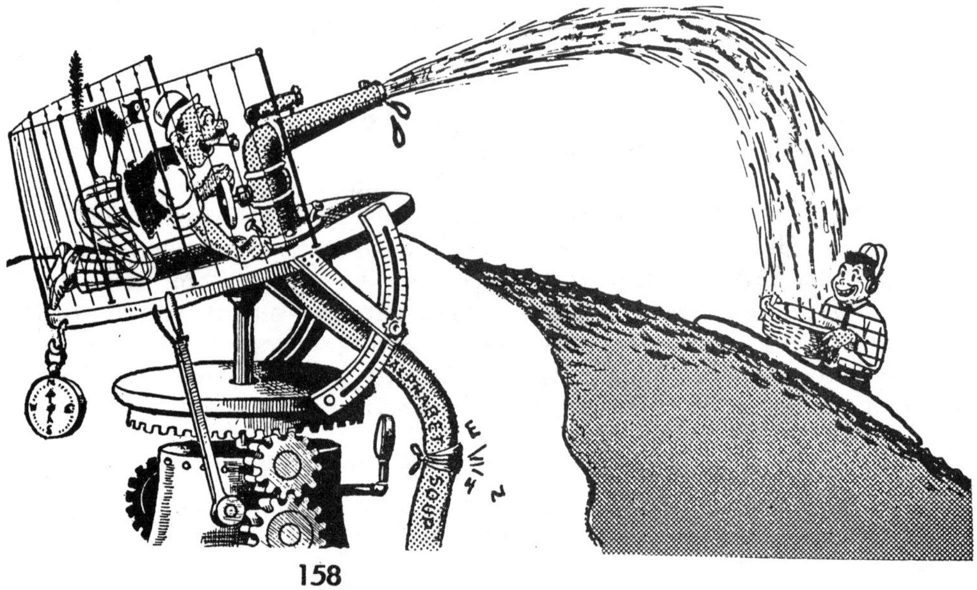

Pick your direction and aim that signal directly at those stations!

The "near" end of the antenna is attached to the side of the building that houses your radio equipment and should be the same height as the mounting at the other end in order that the antenna is parallel with the ground. At each of the two end-support locations, place a screw-eye standoff insulator (sold at most TV and electronics parts stores).

The antenna itself can now be constructed, following the diagram for the general idea. Use something like #12 solid copper antenna wire, it usually comes in 100-foot spools. If you have to splice two ends together, always solder the splice.

When the 108 feet 6 inches are measured (the extra inches are to allow for attaching to the insulators), measure exactly 9 feet from one end and cut the wire. This is the feed point, and you will insert an insulator there. Coaxial center insulators are generally available from well-stocked ham dealers-- one reliable mailorder source has been Van Gorden Engineering, P.O. Box 21305, South Euclid, OH 44121. The center conductor of the coaxial cable lead from your rig goes to the long side of the antenna, the braided shield connects to the short side.

Now you can solder the center insulator into place; also solder the 72-ohm coaxial cable at the feed point-- and do **not** use the 52-ohm type cable normally used for CB installations, this should be TV type cable like RG-59/U. Standard strain insulators are now added to the two outside ends of the antenna.

Hang the antenna on the screw-eye standoff insulators. Use some heavy test fiberglass fishing line for best results, running several strands of thickness for extra strength.

You are now set up for flat side operation-- and you'll about double your signal with the antenna. You could triple your signal by cutting the long side of the antenna to 165 feet 6 inches, although the radiation pattern will change to bring the signal lobes in at closer angles to the end points of the antenna. Want still more power? You can almost quadruple (X4) your signal if you can run the long end of the antenna out to 231 feet 9 inches. The signal will then be booming straight off the ends of the antenna.

The Meter Pinner

The Meter Pinner antenna, in its most evolved version, can give you a 10 dB gain (that's 120 watts PEP for a sideband station) and really punch your signal through. Again, you'll have to figure out in advance in what direction you want your signal to hit, so check your map and compass. You'll also need about 300 feet of space for this monster. It's big, but it's cheap.

Here's how to position the antenna. Locate your base station on the map and place an X there. Next, locate the general target area for your signal and put another X there. Draw a line between the X's, and compute the compass direction. Let's say it's north-northeast (NNE). Take your compass outside and find this direction, placing a marker flat on the ground pointing towards the desired direction.

The antenna itself consists of two horizontal longwires that will straddle this marker. They are placed in a V-shaped configuration, with a 35° angle at the apex (the feed point). The proper angle may be easily determined with a drafting or geometry compass.

The apex of the V, which is the feed point is positioned near the window of your radio room, and a 2X4 plank is used as a mast. Paint the bottom with creosote or something similar to prevent rot, then sink it a few feet into the ground. You'll probably require two more similar masts, one for each end of the antenna's legs.

Measure off at least 291 feet for the far ends of the antenna's two legs, they will be about 175 feet apart from one another at the far end. (If you build one of the miniversions of this antenna, described in the following

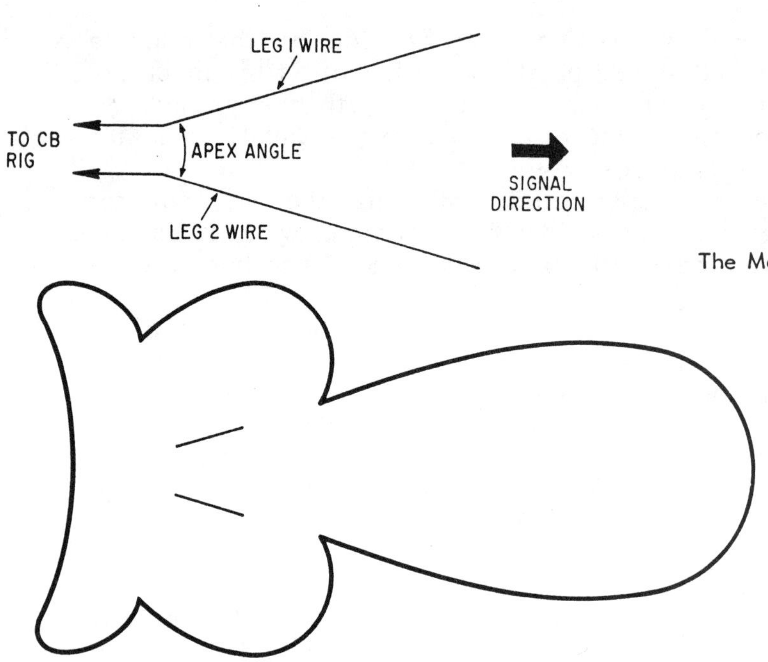

The Meter Pinner: typical signal pattern.

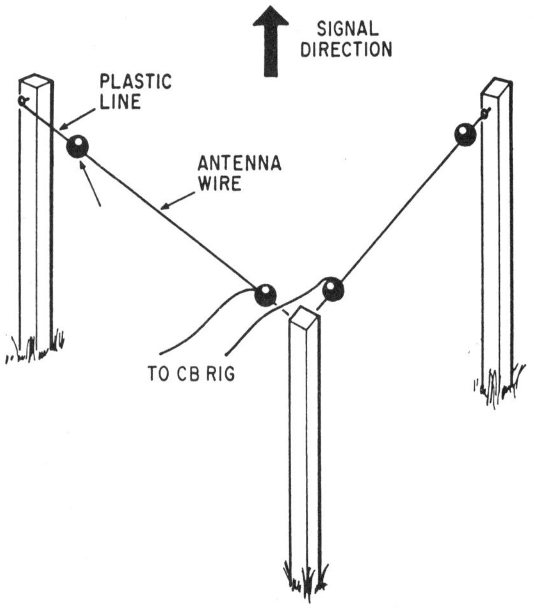

The Meter Pinner: typical construction.

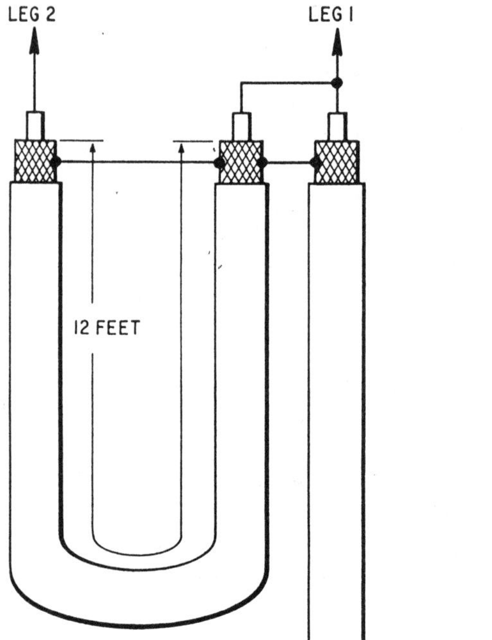

The Meter Pinner: impedance-matching balun.

TABLE

Leg length (ft)	Apex angle (degrees)	Approx. gain (dB)	Approx. effective SSB signal (watts)
217.8	40	9	95
181.5	45	8	76
145.2	50	7	67
108.5	58	5+	38
71.8	70	4+	30
35.4	90	3+	25

paragraphs, the measurements will be reduced.) Erect your far-end masts at these points.

Now that you have established the location of the antenna, the next step is to start hanging wire. Place screw-eye insulators at each of the three antenna termination points. At the feed point, hang two antenna strain insulators from the termination points. At the feed point, hanf two antenna strain insulators from the screw-eye insulator, attaching exactly 209 feet 9 inches of copper antenna wire to each. The actual length of each leg is a few inches longer, however; the extra wire is allowed for connections. Run each of these wires out to its respective far point, and hitch it to some insulators, and from there, to screw-eyes on the masts.

That's all there is to do except to connect the antenna to the rig. It will be necessary to use a matching device, such as a "balun" to make this connection; details are given or you can buy one-- a source is Van Gorden Engineering mentioned in the information on building the Signal Sender antenna. Note that the balun has an input from the CB rig and output connections for each leg of the antenna. A homemade balun should be constructed from 52-ohm (RG-58/U) coaxial cable, wired as shown, and with the braids lashed together as indicated. This balun is to be located directly at the feed point of the antenna, and any length of 72-ohm coax may then be used to run between the balun and the CB rig. Connect the antenna to the rig and you're in business.

If you don't have the space to string up such an impressive brute, make a smaller version (which will offer lower gain). The measurements are shown in the table-- and they do not take into account the couple of extra inches

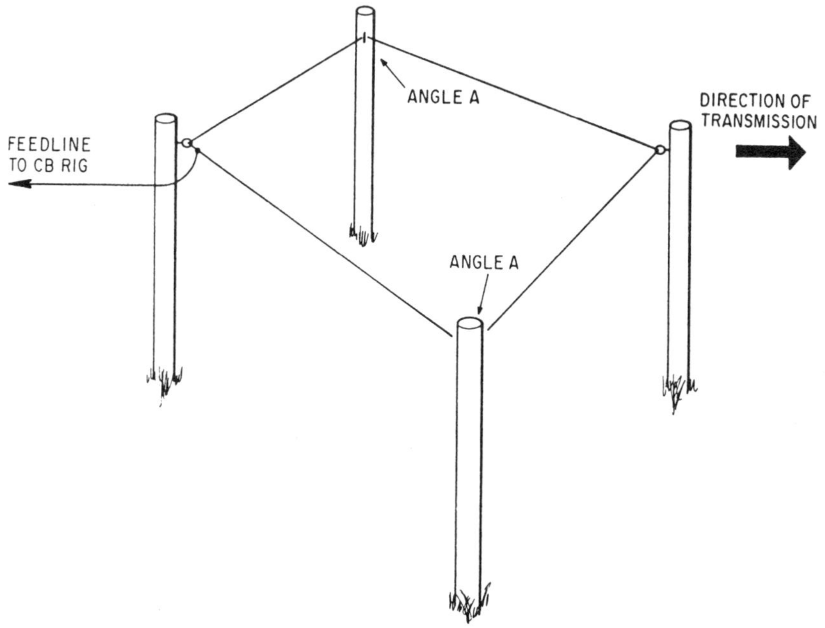

Rhombic: typical construction.

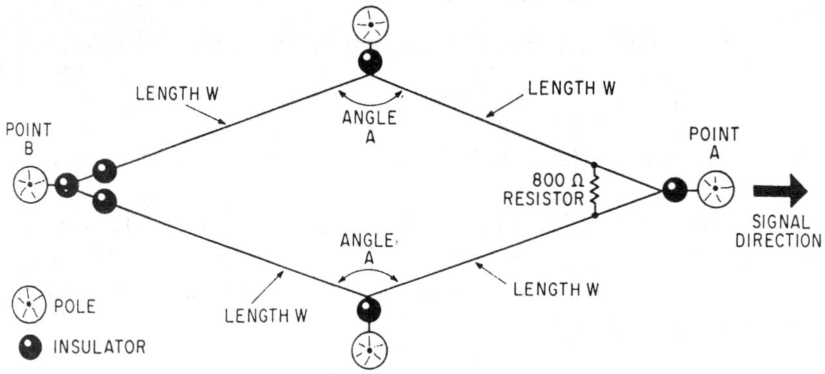

Rhombic:
construction details.

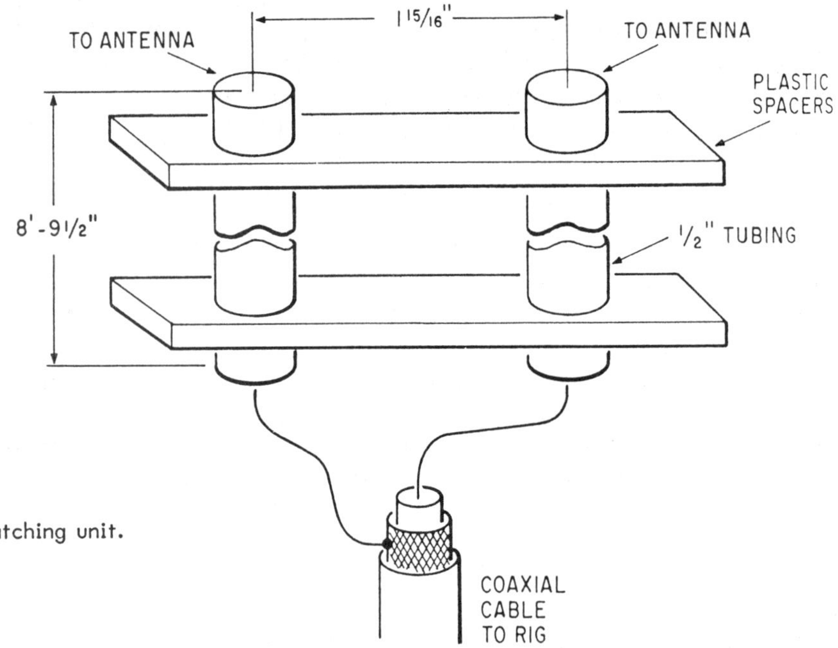

Rhombic:
impedance matching unit.

you'll want to leave for connecting to the insulators. I think you'll like the results.

The Rhombic

The final flat side experimenting antenna is a rhombic type. It's a little more complex than the other two, but (in the largest version shown here) can legally boost a 12 watt PEP sideband signal to one that sounds like more than 200 watts! Of course, you'll need a king-sized piece of turf upon which to put this thing, but it's worth the trouble.

A rhombic will let you drop a most respectable signal in a specific direction, cutting out interference to/from all directions except the one you have selected. Problem is, you can't rotate the thing, so once you've selected your favorite direction-- that's "it." Therefore, this might best be used as an alternate antenna that you can switch over to from an omnidirectional type for general communications purposes.

Ideally, a rhombic should be mounted one wavelength above ground, but FCC CB regulations don't permit this one to be mounted more than 20 feet above the ground.

TABLE			
Length W (ft)	Angle A (degrees)	Approx. gain (dB)	Approx. effective SSB signal (watts)
63.0	70.0	4+	30
126.5	50.0	6	48
252.0	35.0	9+	100
378.0	22.5	12+	200+

The flat side rhombic shown in the diagram is lozenge-shaped with a terminating load resistor installed at one point. The resistor is an 800-ohm carbon-core, noninductive 5-watt type. The resistor is important in order to keep the signal from floating off the back end of the antenna. The resistor is installed at point A in the diagram. Be certain to use only a noninductive type resistor.

Point B in the diagram is the feed point, and you must match your 52-ohm coaxial cable properly at this spot to the 800-ohm load of this particular antenna design. What is required is something called a Q-matching section, made from 1/2-inch OD aluminum or copper tubing, held apart by plastic spacers 1-15/16 inches, center-to-center. The Q-matching section is 8 feet 9-1/2 inches long from the top where it attaches to the feed end of the antenna, to the base where the RG-58/U or RG-8/U coaxial cable from the CB rig is located. The Q-matching section transforms your feed impedance to 800 ohms, matching it to the needs of the antenna.

Length W in the diagram is the same on each of the four legs of the antenna. A variation in size/gain is possible, based upon specs shown in the chart given.

Best bet is to mount this antenna on wooden supports as described for the other flat side antennas discussed. Ise heavy-duty egg (type) insulators, placed as indicated, to keep the wire away from the pole. Nylon rope, or fiberglass, running through pulleys mounted at the top of the four poles, can attach to the back of each insulator. This allows you to erect the antenna at ground level and then hoist it into place over the pulleys with the nylon rope. The rope/pulley combination will also aid you in leveling off the antenna and keeping it on a horizontal plane. You might suspend small bags filled with sand from the bottom of the nylon ropes. This allows the wires to "give" in strong winds-- the antenna will remain taut, pulling on the ropes. The sandbags lift and permit the wire to move in the wind, but lower again when the wind subsides.

Use #12 or #14 copper-clad steel wire for the actual construction. The steel adds strength, while the copper provides good conductivity.

This antenna should offer good coverage towards the direction you select. Two Sidebanders using these at each end of a contact, should be able to regularly communicate via groundwave over a 140+ mile path.

As you can see, by the shrewd combination of the maximum allowable power (sideband's 12 watts PEP) and an antenna offering lots of gain, you can really do a lot.Some commercial beam antennas can deliver better than 14 dB of gain, and that's a serious signal. As a guide in helping you evaluate how a 12 watt PEP sideband signal will emerge when fed into various antennas having different gain factors, a table is shown that indicates the gain of an antenna (in dB) versus a 12 watt PEP sideband signal.

TABLE			
Gain (dB)	Effective SSB signal (watts)	Gain (dB)	Effective SSB signal (watts)
2	19	9	95
3	24	10	120
4	29	11	151
5	38	12	190
6	48	13	239
7	67	14	303
8	76	15	379

13

Freebanders, Outbanders, & Miscellaneous Illegals

Call them Channel Jumpers, Pirates, Bootleggers, Freebanders, Illegals, or any other name you like-- although many folks refer to them by the name I invented for them in the August '79 issue of S9 Magazine, Outbanders. Whatever you call them, they're people who operate beyond the authorized frequency limits of the CB band (26.965 to 27.405 MHz). These unauthorized frequencies have been usurped by Outbanders since the early days of CB and, even though nobody seems willing to discuss them in print, I felt that the amount of activity there calls out for coverage in this book-- for certainly at least 80% of those who operate there are erstwhile CB'ers, although some are licensed hams who have chosen to enter this strange shadow world.

Outbanders proliferate from a starting point at about 26.500 MHz and running upwards in frequency towards CB Channel 1-- this area has been preferred by those using AM communications. The Outbanders that prefer to use sideband prefer the frequencies starting just above CB Channel 40 (27.405 MHz) and heading towards the 28 MHz ham band. Some of the frequencies

On this and the facing page are QSL's from several operators whose ID numbers include ones from groups whose members have been regularly noted on Outbander frequencies.

used by American/Canadian Outbanders are actually authorized for legal CB use in other areas of the world.

Some Outbander frequencies have been especially popular, including 26.800, 26.835, 26.865, 26.935, 27.555, 27.585, 27.890 and 27.945 MHz. Hundreds of stations in the West Indies seem to pop up on 27.835 and 27.845 MHz when the skip conditions are right. At times, these unauthorized frequencies are more populated than the 28 MHz ham band.

No doubt about the fact that all such operations are strictly illegal under existing regulations since those frequencies are set aside for others, including the federal government, industrial stations, and even the CAP. Frequency 26.800 MHz is actually a U.S. military frequency, and 26.620 MHz belongs to the Civil Air Patrol, while 26.650 and 26.770 are reserved for the U.S. Army. The U.S. Navy uses 26.550, 26.560, 26.610, and 26.750 MHz. Rumor has it that 26.760 MHz is used by the CIA!

Above CB Channel 40, 27.430, 27.450, 27.470, 27.490, 27.510, and 27.530 MHz are set aside for industrial stations. Running between 27.405 and 28 MHz

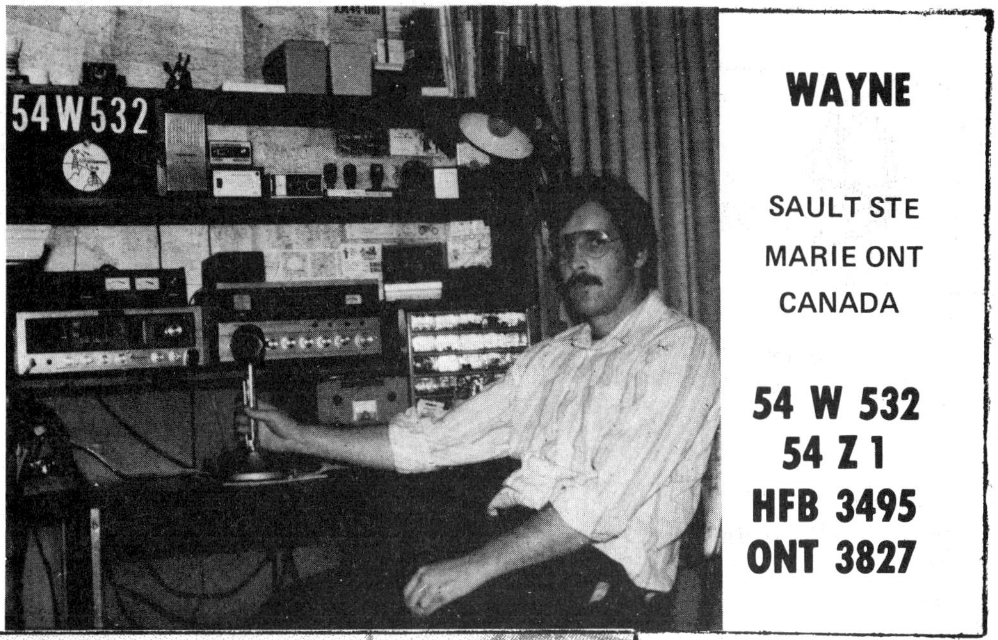
54W532
WAYNE
SAULT STE
MARIE ONT
CANADA
54 W 532
54 Z 1
HFB 3495
ONT 3827

OUTLAW RADIO CLUB 019
PACIFICA, CALIF.

AROI ON 27.685 MHz Ch. 68 LSB
This is ARO-INT'L Calling CQ DX
Music Politics Culture
rebel radio stations

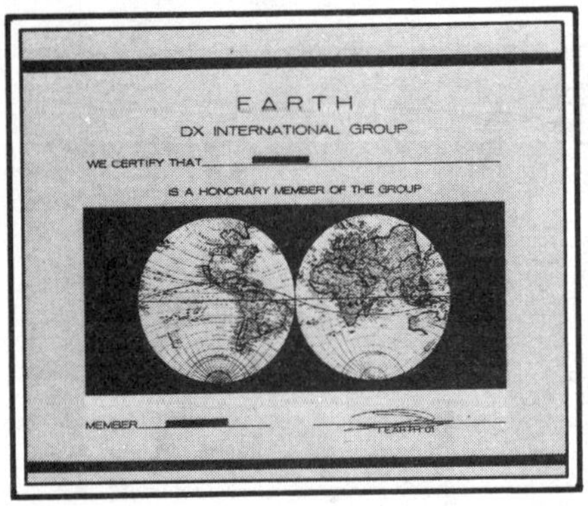

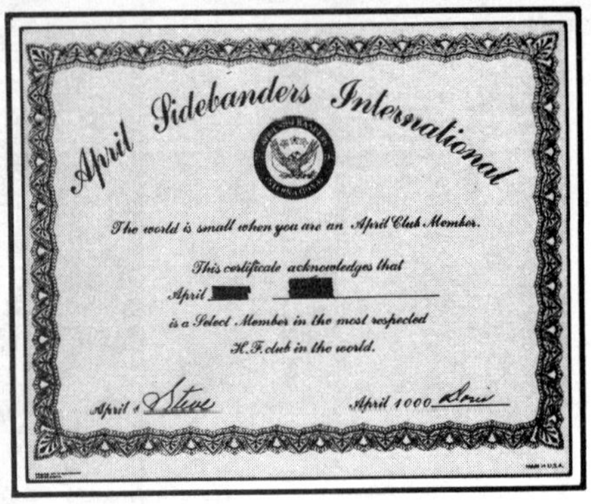

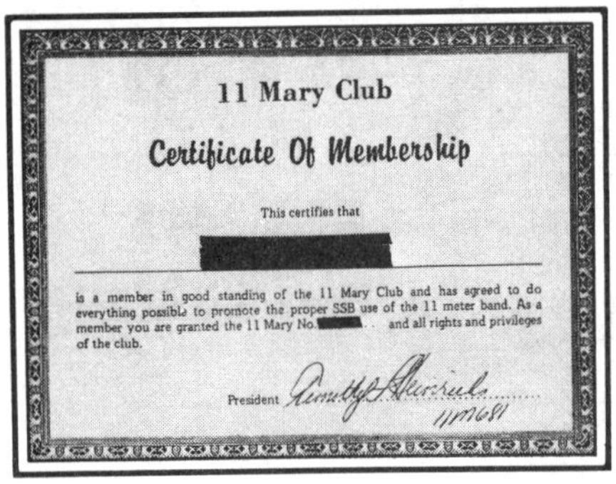

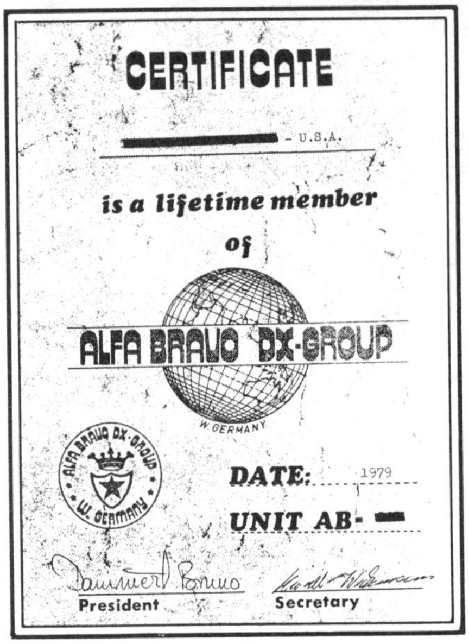

Membership certificates in some groups whose members are sometimes monitored above CB Channel 40. The certificate from the Transcontinental Network specifies 27.870 MHz as the group's frequency. The Alfa Bravo DX Group's certificate is impressively printed on heavy parchment.

is a band that is reserved for federal government communications. Frequencies 27.575 and 27.585 MHz are authorized for use by many agencies, and 27.980 MHz is a U.S. Coast Guard channel. Other frequencies are similarly reserved-- our listing here is just a sampling to give you some idea of typical usage.

As to why there are Outbanders exercising squatters rights in these bands is anybody's guess. Since stations were there long before the CB boom of the mid-1970's, the old excuse "to get away from crowded CB channels" doesn't ring true. Personally, I suspect that it's no more than they way some folks get their kicks-- doing something that isn't allowed. As it turns out, there haven't been any reports of actual harm caused by Outbander operations, and most of the Outbanders I've met seem to be good folks. The major harm done appears, thusfar, to the FCC's ego and ability to cause the Outbanders to vacate the frequencies in order to comply with communications regulations.

If and when an Outbander is caught, the FCC sanctions have been very harsh-- as with any whose operations appear to the FCC to be in repeated and deliberate violation of FCC regulations. The FCC seldom accepts any excuses an operator can come up with. All Outbanders seem to realize this but they take their chances anyway. What with the number of Outbanders on the air, plus their relative anonymity, and the FCC's limited manpower, the chances of actually being nailed aren't especially high unless an operator creates such overwhelming and specific interference problem that a formal complaint is filed by the FCC against a specific station.

In addition to direct monitoring, the FCC seems to have been involved in something of a "sting" type of operation against those whom the agency suspects of dealing in the equipment (export transceivers, transceiver modifications, linear amplifiers, etc.) used by Outbanders. One such dealer told me that he received a letter, supposedly from a Virginia-based CB'er with

THE TRANSCONTINENTAL NETWORK
P.O. Box ▬ · Lawrence, N.Y. 11559

Roster Update

TCN #	NAME	LOCATION	ZONE	TCN #	NAME	LOCATION	ZON
26	Emmett	California	6	568	Dennis	Washington	7
27	Fritz	New Jersey	2	569	Pete	Florida	4

To all 11-M number holders:

1. All 11-M numbers will be given out by personal contact only. (Mail - Telephone - Eyeball etc.). None over the air.
2. The name and address of each new number holder must by forwarded to the person you received the number from. This is to facilitate the distribution of new call sheets and other pertinent material from time to time.
3. If you do not have the capabilities already installed, you are not elgible for a number, until this is done.
4. Numbers will be given out by 11-M number 1 through 11-M number 1C only. No one else has authority to give numbers.
5. You will make absolutely no reference to these numbers or frequency on any other channel or frequency. You will simply suggest "Joe, let's move to 76". Understand? After you once get up there, be prudent.
6. Each new number holder will be given a frequency sheet. You are expected to respect the division of frequencies as outlined on the freq. sheet. Stay away from 10 meters.
7. Lets each of us police our selves. We don't want the same mess they now have on other frequencies.
8. We are all intelligent people, and a book full of rules are not necessary. Please use your good judgement, and we should have a nice place to talk.
9. In case of impending trouble, please let 11M-1 or any of the first 10 number holders know, so that the word can be put out to all. We don't want trouble, just satisfaction, and a quite frequency. OK?
10. This system will start on December 3, 1972.
11. Please destroy this letter after reading it. Under no circumstances, let anyone copy any of the three pieces of information that is furnished to you, except to some one whom you know for sure is a member of this group.

To All:

Good modulating on a brand new frequency.

THE TRANSCONTINENTAL NETWORK

JANUARY, 1980

Welcome to 1980, and welcome to all the new members of the Net. We hope that the group has learned something from the past events of 1979. The Network operation was as successful as any radio net anywhere, on any band. The emergency messages were passed in record time and response relayed through various states and countries to the waiting parties. That's the name of the game.

On the other hand, the Network was jammed by an operator whose toes were stepped on. For almost two weeks the Network was brought to almost a complete halt because someone didn't think. REMEMBER, we hold and operate .870 only through the courtesy of every other operator on the band. Only through propriety, courtesy and clean operation will we be able to maintain our Network. A polite request to a station calling a general C.Q. to Q.S.Y. or if there are stations already in Q.S.O. please give them a chance to break or finish before you request Q.S.Y. In all instances, explain the Net operation then don't just sit and ratchet jaw, request units on frequency to check in or for Q.S.P. for your area. make it sound like a net operation (even if you have to fake the responses). After one or two minutes most units will move elsewhere with no hard feelings.

When moving to an Alt. Freq. or any other freq. **DO NOT ID. WITH T.C.N. OR ZONE.** In this way, if you have inadvertently interfered with another Q.S.O. or group, retribution will not be forthcoming on .870.

With no thanks to the U.S. Postal service and other Postal services around the world, we have managed to get about 75% of the necessary paperwork where it was supposed to go. If there is anyone who did not receive his or her mail don't request it on the air. Our memory is worse than the mail. Please rewrite the request. We will do our best to see that everyone gets everything.

The Net is filling rapidly and the coverage is broadening. We now cover 47 of the 50 United States and 26 areas outside the U.S. There is just about no place on earth outside the reach of T.C.N. and quite a few operators are coming to us from other bands to get their messages through. Take pride in the fact that **YOU ARE GETTING IT DONE.**

Nothing left to say now except for a few quickies.

Watch the quick key.

Avoid .890, that group does not appreciate T.C.N. using their Freq. as an alternate.

The New York P.O. Box can be used as an internet Q.S.L. bureau. Enclose Q.S.L. in a stamped envelope within another envelope addressed to the appropriate unit c/o the N.Y. Box. The Net will forward.

In closing, the T.C.N. wants to say to all members "Congratulations on a successful operation."

THE TRANSCONTINENTAL NETWORK
"All Indians, No Chiefs."

Some of the literature furnished to members of some Outbander groups, including (lower left) an instruction sheet that its senders specified as being one of three restricted documents to be destroyed after reading.

the CB handle **Virginia Sandman.** The Virginia Beach operator asked for information leading to the possible purchase of out-of-band rigs, presumably for resale to his own customers. The information requested was sent and, for whatever reason, not long after the information was provided, the dealer received a notice of monetary forfeiture from the FCC-- the information that had been provided to the **Virginia Sandman** was "Exhibit A" in the FCC "case" against the dealer.

Outbanders, most of whom seem to be primarily interested in shooting skip rather than making local contacts, have tended to view these frequencies with rather well-defined subdivisions. The frequencies below CB Channel 1 are usually referred to as being "LF" (Low Frequency). From 27.405 to 27.585 MHz is generally called "HF" (High Frequency); from 27.585 to 27.755 MHz is known as "VHF" (Very High Frequency); with 27.755 to 27.995 MHz often called either "UF" or "UHF" (Ultra High Frequency). Each "band" has its enthusiastic supporters.

Over the years, the FCC has been especially annoyed with a number of clubs and other organizations that have come along with what the FCC has claimed (correctly or incorrectly) have had the intention of organizing and encouraging Outbanders-- even issuing them on-the-air ID numbers, operating "rules," station rosters, etc. The FCC considers groups that cater to operators on Outband frequencies as being involved in "an organized scheme to operate

radio transmitting equipment illegally." One such group was accused of encouraging operation on unauthorized frequencies." Another group was described in FCC public notices as having their "members operate illegal radio equipment on unauthorized frequencies to avoid detection...except through time-consuming direction-finding techniques."

One group hassled by the FCC was known as HF International, another group was a statewide organization affiliated with a larger group generally known as the Statewide Whiskey group. HF International's members had ID numbers consisting of the letters HF, HFA, HFB, HFC, etc. followed by 1 to 4 digits. The Statewide Whiskey stations had ID's similar in many ways to the old-fashioned FCC CB callsigns, such as 2W1877, 18W8755, 7W6490, etc., except that the numeral before the W ("Whiskey") could be anything from 1 through 99 and signified the state or nation in which the station was located.

Over the years some of the ID's monitored as being used by Outbanders include those presumably issued by groups known as LF International, VHF World Sidebanders, UHF International, HF International, The Transcontinental Network (TCN), Alfa-Bravo DX Group, April Sidebanders International, Earth International, the 11-Mary group, the 11-Michael group, and various Statewide Whiskey groups--among many others. The present status of these groups isn't known. On LF frequencies, there have been groups of operators established that are organized like families, with member stations all having identifications consisting of a distinctive first name plus the family "name." This is instead of the usual type of number-type group ID. So, for instance, the McCoy "family" network might consist of a hundred or more stations, each identifying along the lines of: Junior McCoy, Grandpa McCoy, Effie McCoy, Ezra McCoy, Billy Bob McCoy, Willie McCoy, etc. Some of these families hold huge gatherings or reunions that attract "relatives" from across the continent.

Contacting any such groups for membership is touch-and-go. Most are either semi-underground or else completely underground. The usual way to go about achieving membership requires attempting to establish radio contact with a member and asking how to become a member of the fold. Often it's a Catch 22 situation, since members of some groups seem reluctant or totally disinterested in communicating with those who aren't already members. Inasmuch as the FCC is so down on what it considers Outbander groups, one might wish to weigh the wisdom of becoming a member of same.

Linear Amplifiers

The world is filled with people who want to be better than the next guy. In the world of CB, you can get a more potent legal signal by using sideband instead of AM, and you can use an antenna that offers the benefits of signal-multiplying "gain." Some CB'ers take a different, and unauthorized, approach. They use linear amplifiers.

For those who aren't familiar with the breed, a linear amplifier is a piece of electronic hardware that attaches between the CB rig and the antenna.

Depending on the potency of the linear, the unit has the ability to take the CB signal and amplify it to 50 watts, 100 watts, 250 watts, 750 watts, 1,000 watts, and more! They're known in the CB world under many nicknames such as footwarmers, shoes, sneakers, boots, black boxes, feet, afterburners, socks, tigers in the tank, ankle burners, moccasins, persuaders, hamburger helpers, wearing socks, kickers, or simply linears. They've been around since the very earliest days of CB radio.

Many years passed as thousands of linears were openly sold and advertised to the CB community. Seemingly, so long as the manufacturer included in the advertising a notice that the device was "illegal for use on 11 meters in the U.S.A.," there wasn't any problem in selling them to CB'ers. They weren't illegal to manufacture, sell, or purchase-- it was only illegal to use them.

The early linears that CB'ers used were ones originally intended for ham radio use on the 28 MHz (10 meter) band. They were easily tuned up on or near CB frequencies. These included the Globe LA-1 (200 watts), P&H Electronics LA-400B (200 watts), the Vantron 300 (100 watts), Central Electronics 600L (300 watts), and the imposing Johnson Viking Thunderbolt (750 watts). It didn't take long before companies began producing linears designed especially for the CB market, including the e.c.i. Courier ML-100 (ostensibly made for the "business band"), the super-ugly Elenco L-200, and the shoddy Sonar Radio BR-21. These put out from 14 to 200 watts.

By 1974 the FCC, after being annoyed about linears on CB, came up with new regulations intended to end the problem. The new rules prohibited in no uncertain terms the manufacture and sale of linears intended for use on any single frequency band if that band was located between 24 and 35 MHz. The Federal Trade Commission was even called into the act, lest there be manufacturers/dealers who were inclined to ignore the FCC.

That drove the industry underground, with CB linears still being produced and sold "under the counter" by many suppliers. Some of the linears that were being made prior to and/or after the 1974 "ban" included the Westcom TXL-100B, Pride 150, and models from Varmint, Hooker, S.T.E., Telco, Maco, Marko, D&A, Stacom, Palomar, Digital Sport Systems, Thunderbolt, and Norcom-- plus an endless number of anonymous "junker" units that have no name inscribed thereupon. Many are homebrewed in garage and basement "factories," and are quite seedy in appearance, design, workmanship, and performance. The 100-watt mobile linears from such sources are especially cheesy.

Some linears are known as being "bi-linear," that is, they are equipped with circuits that boost the received signals in addition to the transmitted signals. The FCC doesn't care about the received signals being boosted, it's the transmitter signals that they don't want souped-up.

The FCC claims that linears create interference to the communications of CB'ers and to TV reception due to the powerful signals overloading the front ends of such receivers. Linears that are poorly tuned or badly designed/made can actually generate a multitude signals on many harmonic and spurious frequencies that can raise havoc for miles around and adversely effect police, fire, broadcast, and navigational systems.

It's been my experience that there are better ways to improve 27 MHz communications than by the use of a linear. Moreover, a station running a heavyweight linear is also more likely to be breaking a number of FCC regulations such as being an Outbander or shooting skip. Should an FCC monitor be tuning across the band looking for a likely candidate for a hassle, the clown with the macho linear will surely be the one that stands out best and says "Nail me first!"

Many ham transceivers have been modified by Outbanders for their own purposes, however none ever seemed to become as sought-after as the legendary Yaesu FT-101 series. Even though the units were made in the 1970's, they still command top dollar. It runs 80 watts AM and 260 watts PEP on SSB.

Ham Equipment on CB on Outbander Frequencies

Ham transceivers designed for multi-band use being used by CB'ers and Outbanders are nothing new. Certainly, at least half the production run of Yaesu's wonderful FT-101 series of AM/SSB transceivers seemed to eventually find their way into the hands of CB'ers and Outbanders. The set came out in the 1970's and still ranks high on the popularity list with its 80 watts of AM and 260 watt PEP SSB capabilities. Operators quickly learned to pull out the unit's "10A," "10B," and "10C" 10-meter ham band crystals and substitute new ones for LF, CB, HF, VHF, and UHF operation. A 32.52 MHz crystal in the "10A" crystal socket bestowed operation from 26.500 through 27.000 MHz; a 33.02 MHz crystal provides 27.000 through 27.500 MHz coverage; and a 33.52 MHz crystal in the "10C" socket provides 27.500 through 28.000 MHz coverage. These old sets still bring from $500 to $600 in good used condition-- even though they are patently unauthorized on any frequencies other than the 10/15/20/40/80/160 meter ham bands by duly licensed hams.

Of course, ever since the FT-101 became so popular, many other newer ham transceivers have come along from various manufacturers and have been taken to the bosoms of persons interested in their use on CB and Outbanding frequencies. Some require a minimal modification, others don't require any at all-- although the prices of new transceivers can run between $900 and $1,200. What can I say, if you get caught using any unauthorized equipment you are at the mercy of the FCC, which seems to view such antics with a coldly dispassionate eye. Chances are, you'll get hit with a "notice of apparent monetary forfeiture." That means, they say you owe them a certain amount of money (like maybe $2,000). You can appeal it, but your appeal isn't heard in a court of law, but by an FCC staffer that the agency has dubbed an "Administrative Law Judge." The cat's not a real judge, just a civil service employee on the FCC's payroll. Fat chance of talking your way out of paying at least a portion of the original bite, even if it's your first offense.

CB Broadcast Pirates

You know about pirate broadcasters in the AM, FM, and shortwave bands, but those that have operated on the CB channels are there too, despite lesser notice by the news media.

Undoubtedly the desire to be a broadcaster, combined with the ease of putting a 27 MHz station on the air offers the opportunity to those who have gone this route. Actually, although it's a strange notion to take, it probably isn't any more an unauthorized use of the CB channels than shooting skip, running a linear amplifier, and other more commonly encountered affronts to FCC regulations. The major difference is that CB operators have generally been more irritated than amused with those who have thusfar established such stations-- probably because it results in one less channel for ratchet jawing in a local area; when the skip's in, the broadcaster could knock out any number of distant areas, too.

Although FCC enforcement efforts have always been in evidence on Outbander frequencies, stations have been there for many years. Here, FCC agents inventory some of the booty confiscated during a raid on Outbanders in the Washington, DC area. Looks like they got everything but the coffee cups.

What we are talking about here are lengthy, often regularly scheduled, one-way transmissions of music and other programming material transmitted for the primary purposes of reaching a wide audience for the purposes of entertainment. These aren't your basic "chuckers," those who get their jollies by jamming a channel for a few minutes by keying up their CB mikes while holding them in front of the loudspeaker of an AM or FM receiver-- just for the sake of being annoying. Chuckers are commonplace and represent those who have decided and discovered that CB is more fun and less expensive that going to a shrink.

Nor are the broadcasters who have a sort-of mini-broadcast lasting for 3 to 5 minutes that they send out between skip contacts. A typical example of this is a guy I hear from time to time on Channel 12 (have heard him for years) almost every time the skip opens towards Dixie. His CB handle is **Music Man** and his method of attracting DX contacts consists of a long winded and overly repetitive monologue of his handle, location, and his sincere call to stations around the world to become his next contacts. This information is padded out by a lengthy musical interlude mostly centered around a display of banjo pickin' and hot fiddlin' that sounds tantalizingly like The Ballad of Jed Clampett played at a speed that I can't determine as being slighly too fast or a shade too show-- but certainly one of the two. I believe that the entire station announcement is pre-recorded because it sounds similar each time it is sent out.

Music Man has a method to his apparent madness. He wants to retain complete control of the channel for his own purposes. It does seem to be an effective stunt. Even when the channels are wall-to-wall with skip stations, the only DX activity I can hear on Channel 12 consists of **Music Man** and sometimes the station he's working, plus occasional anguished cries of other operators demanding that he get off the channel.

Music Man isn't alone in this practice, but he's the one I hear most often. Another station I hear from time to time regularly plays the chant of a

tobacco auctioneer. There's another guy with bagpipes, one with a barking dog, someone who plays Sousa marches, another who likes sound effects (bells, whistles, sirens, diesel horns, etc.). Best of all is the guy who sends out a long ear-busting series of hog calls to attract DX stations-- one of the most unusual bits of unintentional humor ever let loose on the band.

To the "true" CB broadcaster, these feeble and brief efforts are minor league. Sometime around 1979 on NBC's Today Show they ran some videotape taken right in the studios of several CB broadcasting stations in California. Since then, stations have popped up in all areas. These broadcasters are on for several hours per day, they have ID's, DJ's and QSL cards.

Large amounts of money have been lavished upon some of those that have come along, with professional studio gear and large record/tape libraries. Telephone numbers are given out for music requests, and one station I heard recently during a skip opening would stand by on the CB channel just in case anybody wanted to send a request by CB radio.

Chances are that the majority of CB broadcasters are far less exotic, consisting of no more than a CB rig, turntable and tape deck. The operators are no less enthusiastic than the fancier installations. Of course, it's still open to question as to whether or not there is actually any audience for what these stations have to offer in the way of entertainment.

But, needed or not, and whether listened to by choice or force, CB pirate broadcasters have been around for years. In fact, in 1978, there were reports of a **Radio Vocad** on 28.865 MHz, right in the 10 meter ham band!

So far as I have been able to learn, thusfar none of these stations have been caught or hassled by the FCC.

14

CB Thefts/Security

Every day sees numerous CB rigs ripped off from cars. Depending upon the installation, the vehicle, the rig, and the neighborhood, a CB rig might well be lifted within hours after it was put into a car. Years ago, rigs were stolen by those who were interested in using them and giving them out to their friends. Then, the emphasis shifted to selling them at bargain prices to strangers met at CB jamborees, coffee breaks and other similar gatherings. That still goes on, especially since more costly rigs usually find ready buyers who are less interested in the ancestry of the unit than the low price. The seller finds it a quick way to get drug money.

A pro can yank a CB rig from a locked car in a few seconds. A good many professionals have sets of keys that can open a locked car as quickly as the vehicle's owner. Bent coat hangers, "Slim Jims," and even a hammer-and-towel combinations do the trick when keys aren't available. Junkies prefer the brute-force method of smashing the window in with a hammer or a rock. Pros would prefer do use a key and do the job with as little noise as possible. Of course, many people don't bother to lock their cars and that makes it easy all around.

CB thieves prefer to work nights. They especially like dark parking lots at at motels and hotels where a goodly percentage of the patrons probably don't have any business being there in the first place (and will therefore probably not report the theft to the police). Vehicles there with out-of-state license plates are also prime targets since their owners are either business people or tourists who don't want to hang around town filling out police forms and waiting for the rig's possible (but unlikely) return. Parking lots at sports events, airports, and movies are also popular ripoff spots. In major cities there are long-term parking lots that accomodate cars for days or

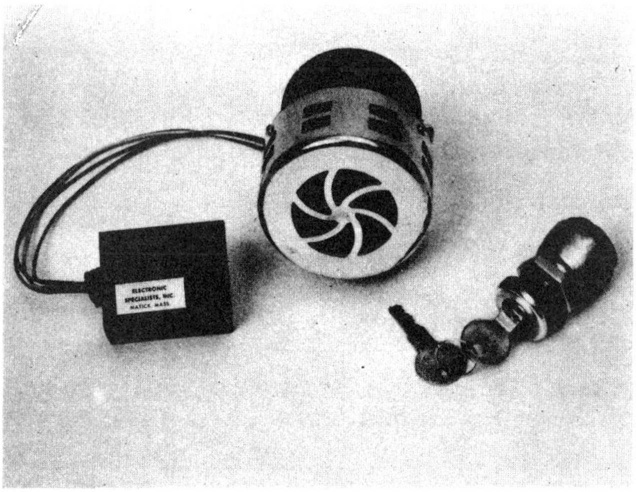

An alarm with a really loud and long-howling siren is much preferred to one that keeps honking the car's horn-- most people don't sense that the car horn they're hearing is part of an alarm system.

After having one CB rig swiped, a Florida operator designed this good looking sheet metal and plywood console to hide and protect his equipment. A little bit of planning and some effort in your basement workshop could put a customized unit like this into any vehicle at minimal cost to the operator.

weeks at a time; those are favorite spots, too. Other high risk parking areas are are the driveways of private homes, alleys, parking lots at factories and industrial complexes, restaurants, and poorly lit streets.

Discos and bars, surprisingly enough, aren't as attractive to pros as some other spots, the crooks fearing the sudden arrival of a nasty drunk or the unhappy discovery of a couple of lovers hidden in the rear of the car.

CB ripoff artists who work days like shopping mall parking lots, museums, factory lots, colleges, tourist attractions, and CB jamborees. It isn't unheard of for parking lot attendants to work in cahoots with the thieves. The snappy CB whip poking up above the vehicle is an arrow pointing the way to the equipment. The time between the mobile unit being spotted and the CB rig being gone is very brief.

Cheaper, older, and off-brand AM units have less reseale value and appeal than newer, major brand, deluxe SSB transceivers. Rigs with a high resale potential are quickly disposed of at CB jamborees and coffee breaks, in flea markets, and from the backs of trucks. A top-of-the line unmodified sideband rig is usually easily sold when it's offered for only $75 to $100. Even junky little El Cheapo or Mickey Mouse AM units can be dumped if offered for $10 to $15 on the juvenile market. Some thieves go so far as to swipe the mobile antenna along with the rig and offer the whole thing as a package deal.

Almost any protection effort (short of removing the rig from the vehicle and taking it with you) can be foiled by a determined pro given a little time. Making it hard is, in itself, a deterrent of sorts because the idea is to rip

A magnetic mount antenna is an excellent security accessory. These antennas function quite well, will stay attached at any speed, and can be instantly removed when you leave the vehicle unattended. If you've got a convertible, or a vehicle with a vinyl covering on the roof, you'll have to place it on the hood or (preferably) the trunk lid.

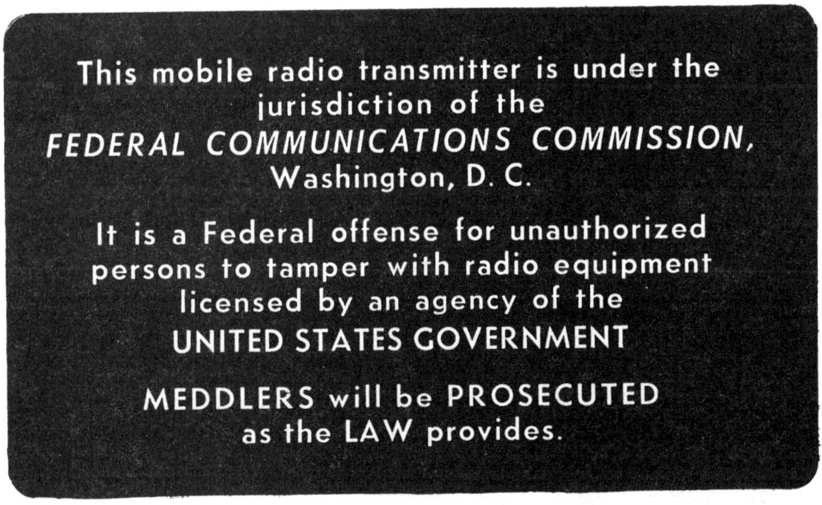

Some CB shops offer an array of different warning decals that are worded in such a way that they may give second thoughts to someone who is eyeballing your mobile CB rig with unwholesome intentions.

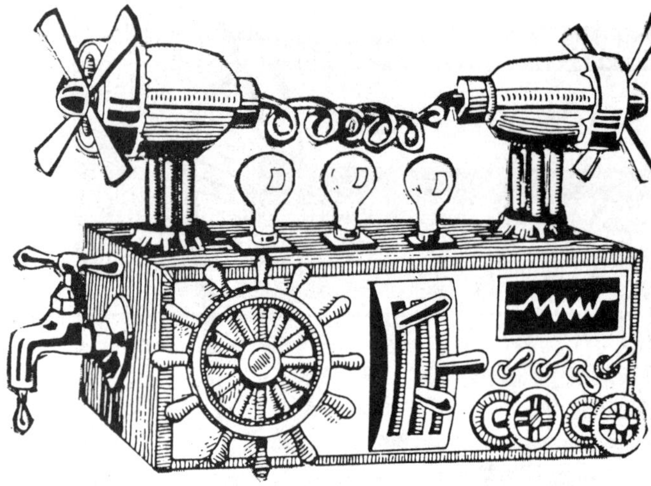

Rigs that have been greatly and obviously modified, or painted some hideous color, are seldom ripped off. On the other hand, they may look rather ugly 'neath your dashboard and you can't ever sell them.

and run-- nobody wants to make a career out of pulling a rig-- a risky business. Some thieves will pass by a rig that looks as if the vehicle owner has made any reasonable effort to protect the unit past basic locking of the vehicle's doors. Rigs that look like they can be taken with the least amount of work and risk are certainly preferred.

The CB thief has several things going for him-- and once you grasp what they are you can then take steps to putting up a good defense. You see, there's a flaw in the human mind that makes most people think that it can't happen to them, "Those things only happen to the other guy." Well, they happen to somebody, why not you?

Also, most people don't bother to record the serial numbers of their rigs and/or keep the sales receipts. If a rig is stolen, they don't have an adequate way of describing the unit to the police, or identifying it if asked. How in the world could you convince the police that a particular "Frammis-40" rig was yours when they call you in to look at a dozen of them they recovered? Chances are that half of the stolen rigs recovered by the police can never be returned to their owners because fow few people could fully identify them-- guys who didn't keep a record of the serial numbers because it was the other guy's rig that was going to be clipped.

Two clever antenna security ideas. At the left, a quick-disconnect adapter that can be used with most whip antenna types. At the right, a hinged mount that sticks out from the vehicle's trunk when in use. When you park, open the trunk and it hides away in a flash.

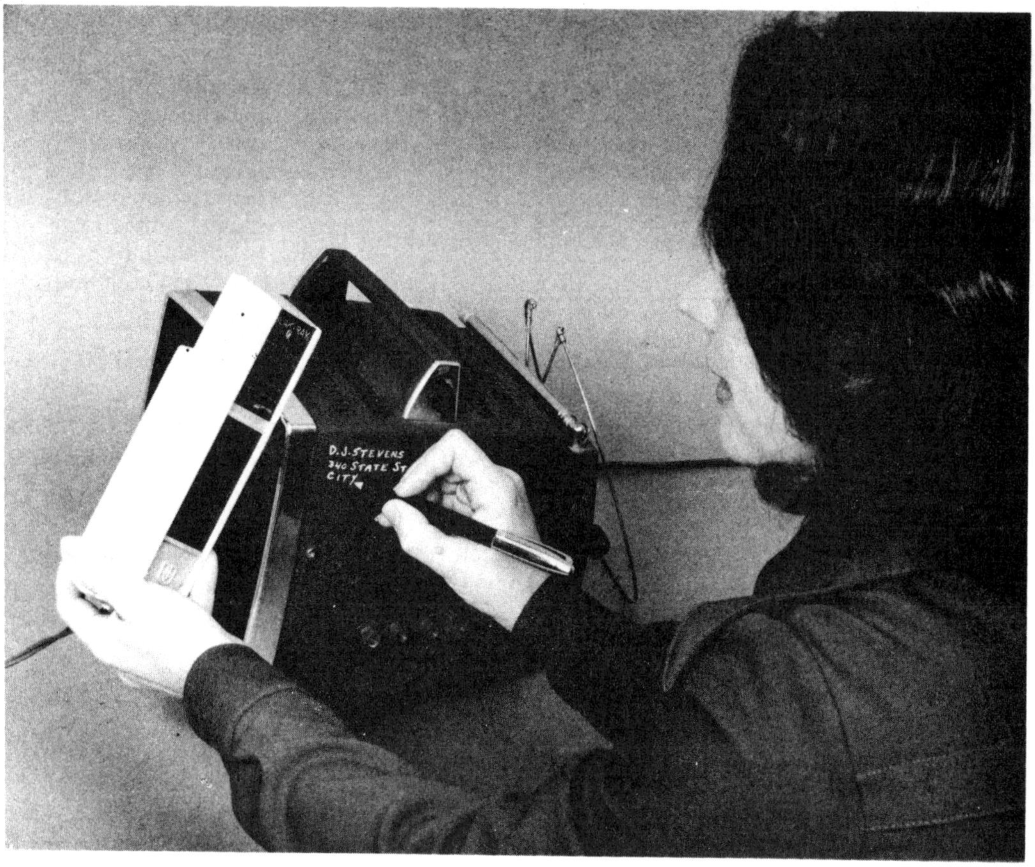

Painting your identification on electronic gear with ink that shows up only in ultraviolet light is an approach that many use.

Removing the rig from the car when you get out is really the safest way to assure theft-protection. Get a sturdy slide-mount, and when you get out of the car, take the set with you in an attache case. Remove the antenna too by means of a magnetic, snap-on, or a whip with a quick-disconnect type of device. There are also hinged mounts that extend out from inside the trunk and can be folded over into the trunk when you park. Check out the possibilities of AM/CB cowl-mount disguise antennas that look like regular car antennas.

If you want to hide your gear in the vehicle's trunk or glovebox, pull over to the side of the road a couple of blocks before you arrive at your destination to stow the rig. What good is it to hide the rig in plain sight of the guy who's looking to seal the thing?

Some CB'ers have hooked up various booby traps such as hidden fish hooks, razor blades, tear gas guns, etc. that are intended to surprise and dissuade would-be CB thieves. This isn't a good idea and is probably more complicated than other steps that might be taken. Also, you could hurt yourself, or some member of your family with any kind of a trap. With the way things have been of late, a crook who got injured on your booby trap could sue you and the police would be looking to lay a concealed weapons charge against you!

One of my good readers suggested that a great way to dissuade CB thieves would be to spray paint the rig and antenna some hideous color such as day-glo fluorescent red, orange, green, or yellow. It would lose its attractiveness and ability to be sold-- but easy to spot should it ever actually get stolen and turn up elsewhere.

The most obvious attempt at protection would be to lock the vehicle with its windows rolled up. And forget about attempting to hide a spare

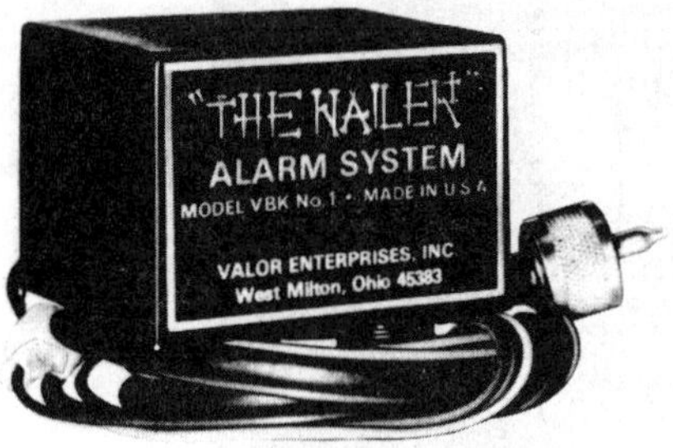

A number of small alarms are made to sound off when the CB rig or antenna are disturbed. The problem is that they don't make nearly enough noise when they go off.

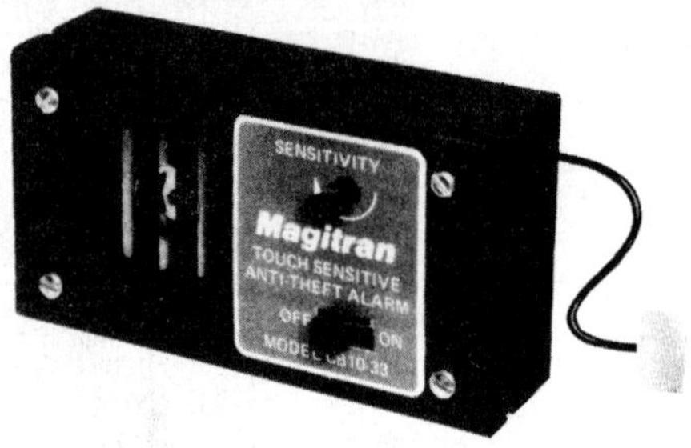

key on the car by means of tape or magnets-- thieves already know to look for such goodies under the fender, in back of the license plate, in back of the bumper, in the engine well, inside the gas cap door, etc. They'll be more than happy to use any key you have so thoughtfully left for them.

Might be a good idea to install any good loud bell or siren type of vehicle burglar alarm system hooked to the doors, trunk, hood, seats, and ignition. Forget about the alarms that blow your car's horn, those who'll hear it won't recognize it as an alarm. Remember, however, that some alarms have a 20-second delay to allow the vehicle's owner to get into the vehicle. That 20 seconds is one-third of the time it takes a pro to pull the rig, so the alarm is actually sounding for only about 40-seconds while the thief is actually in your vehicle. Still, it's better than no alarm.

Of course, you want to let prospective thieves know that the car has an alarm. Alarms usually come with stickers for your windows, be sure to use them. Those fake green/yellow alarm stickers sold in novelty stores don't fool anyone and serve only to assure crooks that the vehicle doesn't have an alarm.

If your vehicle has doorlock buttons with mushroom-shaped tops, replace them with ones that don't have that bump at the top so a bent coathanger can't be used to open the door. They're cheap at any auto supply store.

Park in well-lit areas where there is plenty of pedestrian traffic. Park with the front of the vehicle facing street lights and passersby.

Besides keeping a record of the rig's serial numbers, engrave that number as well as your name and driver's license number (not your address) on the chassis of the rig. Some rigs have the serial number on a flimsy decal or stick-on tape-- this is worthless and is quickly peeled off. A felt-point waterproof black laundry marker can be used to put some of this information (or some secret symbol) anywhere you can room on a printed-circuit board. Yes, these measures might lower the reseale or trade-in value of the rig when you go to get rid of it-- but at least you'll have it to get rid of.

Many CB'ers think that engraving their Social Security number on a rig is of some great benefit. Not so, since the SS Administration won't tell the police whose name matches up with those numbers, although the police can sometimes get a match through armed forces serial number computers.

And, by all means, speak to your insurance agent or company especially in order to find out the extent of any coverage you have on the CB equipment in your vehicle relating to theft (and also accident, while you're at it).

Reports have come through of several theft scams that rather bold thieves have found to be effective. In one version, they cruise around in mobile units much like what the average CB'er could accept as an unmarked FCC mobile unit. They look for in-transit CB'ers, possibly even parking by the side of the road to flag 'em down as the pass or stop for a traffic light. The CB'er is then quickly flashed a meaningless but fancy ID card and badge and told that the agency is doing random CB inspections. The supposed FCC

If a card like this won't scare 'em off, nothing will!

man fiddles with the controls a bit and holds a field strength meter up to the set just long enough to determine that the set is either overpowered, overmodulated, underpowered, undermodulated, off-frequency, splashing over, illegally modified, or generating harmonics-- possibly all of these things. The only solution is for the bogus FCC-man to drag out a clipboard and start writing up all sorts of official-looking greenstamp papers, otherwise (he claims) he could simply confiscate the offending equipment for "further tests and adjustments." Amazingly, lots of CB'ers happily volunteer their equipment for "confiscation" rather than be smacked with what they are given to believe is several hundred dollars worth of fines.

A variation of this gambit doesn't call for equipment confiscation, only that the name and address of the CB'er be recorded by the "FCC man" for his files to show that the rig passed "inspection." If it was an expensive rig, it is invariably swiped from the vehicle within 48 hours.

FCC people (real ones) don't have the authority to direct traffic or affect the operation of motor vehicles. They can't direct a mobile CB'er to pull over to the side of the road-- police officer, sheriff, or state trooper can do that. If anybody tries to do a roadside CB inspection on your rig, jot

At the left, a disguise antenna that does double duty for CB and the car's AM/FM radio. Doesn't look like a CB antenna. To the right, a heavy-duty lock mount for the rig discourages theft of equipment.

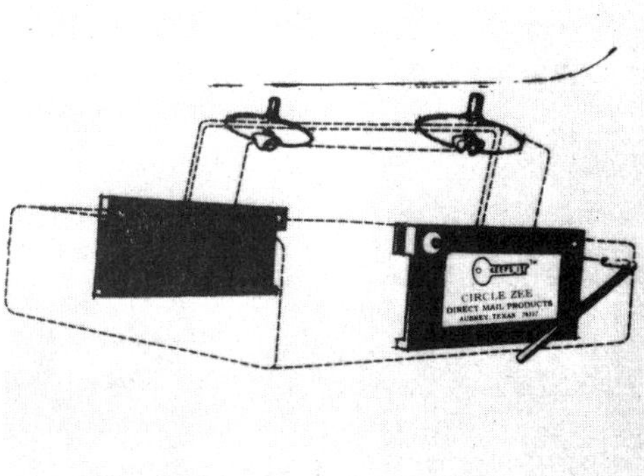

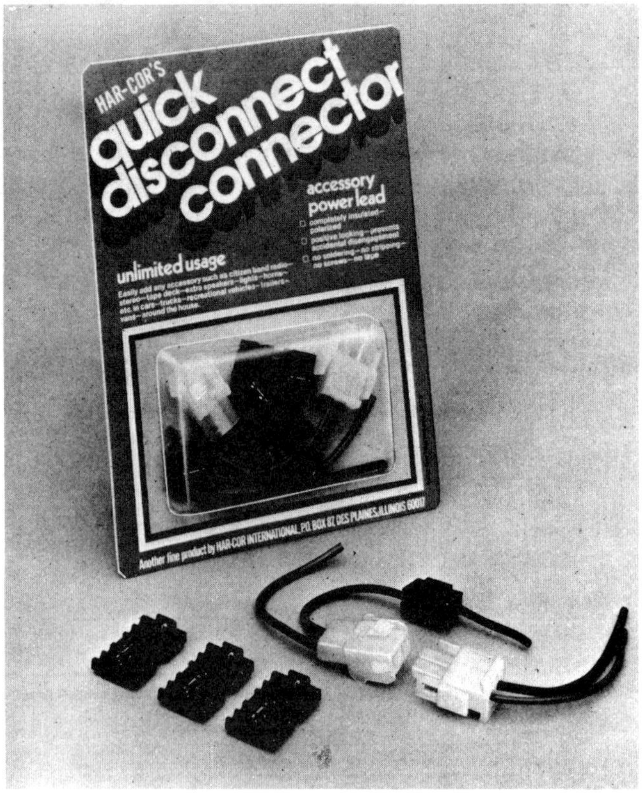

For those who don't wish to go all out with a slide mount, quick-disconnect power connectors are a painless and inexpensive way of making it easy for you to remove your rig from the vehicle when parking.

down their license plate number and a description of the vehicle and report the incident to the nearest FCC field office.

CB'ers have also been relieved of their gear after being summoned to the scene of a disabled female motorist located at a remote spot-- or when they have called for help themselves after running out of gas or having a mechanical problem. CB'ers have had equipment stolen upon opening their locked mobile units in dark parking lots-- so the faster you can get your door opened and then locked, the better. And don't open your car windows until your vehicle is under way. And keep your doors locked and windows rolled up when driving through any known high-crime area-- it takes no more than a traffic light for your vehicle to pick up one or more uninvited passengers. Your rig could be removed and hustled down the crowded streets even before the light turns green again!

Never announce your last name, home address, or telephone number over the air-- use a P.O. Box number if you want to tell someone where to write to you or send a QSL.

The almost casual disregard for the most basic safeguards for CB gear is startling. Many CB's are stolen because their owners were virtually indifferent to the security of the equipment-- leaving many tempting morsels for those who know that the gear can be easily converted to cash.

15

CB Around the World

These days, CB exists throughout the world, and with varying degrees of similarity and difference from what CB is in the USA and Canada. CB equipment, channel allocations, and regulations can be quite different than much of the equipment and many of the FCC regulations shown for the USA generally discussed in this book. Readers in other nations would be well advised to check with their governmental telecommunications authorities regarding CB regulations. Likewise, CB equipment suppliers in the different nations should be contacted regarding the latest equipment meeting local requirements.

Some of the more active CB areas of the world include:

Australia: Australians are avid CB'ers. CB history in Australia went through an early period of illegal 23-channel 27 MHz band radios that eventually became legalized when the government attempted to create an 18-channel HF CB service-- 16 of the 18 channels were from the original 23, and 2 new channels were added in between. Ultimately (and presently), Australia's 27 MHz HF CB service ended up with 40 channels. These are identical in frequency and numbering to the channels in the USA, Canada, and which have now become somewhat standardized in many nations of the world. AM and SSB are allowed.

The way this band is generally used in Australia is with Channel 8 as the Highway Call Channel. Channel 11 is the General Call Channel. Channel 9 is for Emergencies. AM operators usually operate on Channels 1 to 7, 10, and 12. SSB operators are fond of operating between Channels 16 and 40.

More recently, a 40-channel UHF CB band was approved in Australia in addition to the HF band. This useful band runs from 476.425 to 477.400 MHz, with 25 kHz channel spacing, narrow FM mode. Much use is made of repeaters on certain channels. Commercial and private use is allowed, and in some circumstances (such as road transport), good use is made of both HF and UHF bands in conjunction with one another.

In the UHF band, Channels 1 through 4, and 6 through 8, are paired semi-duplex channels as follows: 1/31, 2/32, 3/33, 4/34, 6/36, 7/37, 8/38. Channel 5, which is the Emergency Channel, may be used either simplex or semi-duplex. Channels 9 to 30, plus 39 and 40, are simplex. Channel 40 is the Highway or Transport Operators chit-chat frequency.

Equipment names most familiar to Australian CB operators include Pearce-Simpson, Dick Smith Electronics, GME, Uniden, Midland, and Roadrunner. Australian CB operators like to read CB Action magazine, which has been in operation for many years. The author acknowledges and thanks the cooperation of his friends at Dick Smith Electronics Pty. Ltd. in the preparation of this information.

Australia's UHF CB channels are:

Ch. 1	476.425	Ch. 11	476.675	Ch. 21	476.925	Ch. 31	477.175
Ch. 2	476.450	Ch. 12	476.700	Ch. 22	476.950	Ch. 32	477.200
Ch. 3	476.475	Ch. 13	476.725	Ch. 23	476.975	Ch. 33	477.225
Ch. 4	476.500	Ch. 14	476.750	Ch. 24	477.000	Ch. 34	477.250
Ch. 5	476.525	Ch. 15	476.775	Ch. 25	477.025	Ch. 35	477.275
Ch. 6	476.550	Ch. 16	476.800	Ch. 26	477.050	Ch. 36	477.300
Ch. 7	476.575	Ch. 17	476.825	Ch. 27	477.075	Ch. 37	477.325
Ch. 8	476.600	Ch. 18	476.850	Ch. 28	477.100	Ch. 38	477.350
Ch. 9	476.625	Ch. 19	476.875	Ch. 29	477.125	Ch. 39	477.375
Ch. 10	476.650	Ch. 20	476.900	Ch. 30	477.150	Ch. 40	477.400

Czechoslovakia (former): Has 20 channels of 4 watt AM/SSB/FM. These correspond to standard CB Channels 1, 3, 4, 5, 7, 9, 11, 12, 13, 15, 15, 17, 19, 20, 21, 24, 25, 26, and 27.

Eire: Has the standard 40 channels, allowing 2.5 watts on AM, and 4 watts on FM.

France: Standard 40 channels, 4 watts FM, 1 watt AM, 4 watts PEP for SSB.

Germany: Has the standard 40 channels, allowing 4 watt FM, 12 channels of 1 watt AM, 22 channels of 0,5 watt FM, 12 channels of 0.5 watt AM/FM, and 12 channels of 0.5 watt AM.

Great Britain: In the UK there are CB activities on several bands. There are FM operations allowed on the standard 40 channels, plus FM operations on 40 channels spaced at 10 kHz steps between 27.60125 and 29.99125 MHz (i.e. 27.60125, 27.61125, 27.62125 , etc.). There are also 20 FM channels (50 kHz spacing) situated between 934.025 and 934.975 MHz. Many UK operators can be heard using AM and SSB on 27 MHz, although it does not appear to be permitted.

Greece: Has the standard 40 channels, 5 watts AM.

Hungary: Has the regular 40 channels of AM/SSB/FM. AM/FM units are allowed 4 watts, with SSB units running 12 watts PEP.

Italy: Italy offers the standard 40 channels with 40 watts. There are 34 AM/FM/SSB channels designated for business users.

Netherlands: 40 channels of 2 watt FM, 22 channels of 0.5 watts FM. CB violations (such as AM or SSB operation, and overpower) are heavily enforced by the government.

New Zealand: A very active CB nation! As of January, 1988, the CB service was expanded and is now a full 40 channels, although not the same frequencies used in many other nations. In New Zealand, Channels 1 to 14 are for AM only. Channels 15 to 24 are also for AM, but are designated for Calling/Emergency purposes. Channels 25 to 40 are set aside for SSB operations. We appreciate this information from our friends at Dick Smith Electronics Pty. Ltd., who also gave us the following New Zealand CB channel information:

Ch. 1	26.33	Ch. 11	26.45	Ch. 21	26.58	Ch. 31	26.68
Ch. 2	26.34	Ch. 12	26.47	Ch. 22	26.59	Ch. 32	26.69
Ch. 3	26.35	Ch. 13	26.48	Ch. 23	26.62	Ch. 33	26.70
Ch. 4	26.37	Ch. 14	26.49	Ch. 24	26.60	Ch. 34	26.71
Ch. 5	26.38	Ch. 15	26.50	Ch. 25	26.61	Ch. 35	26.72
Ch. 6	26.39	Ch. 16	26.52	Ch. 26	26.63	Ch. 36	26.73
Ch. 7	26.40	Ch. 17	26.53	Ch. 27	26.64	Ch. 37	26.74
Ch. 8	26.42	Ch. 18	26.54	Ch. 28	26.65	Ch. 38	26.75
Ch. 9	26.43	Ch. 19	26.66	Ch. 29	26.66	Ch. 39	26.76
Ch. 10	26.44	Ch. 20	26.57	Ch. 30	26.67	Ch. 40	26.77

Poland: Poland has 40 channels of 4 watt AM/FM CB, but the channels are 26.960 to 27.400 MHz, which skews them 5 kHz down from the standard 40 channels.

Portugal: Standard 40 channels with 5 watts AM/FM, 15 watts PEP on SSB.

Spain: Same 40 channels and tech standard as USA/Canada.

Switzerland: Has the standard 40 channels, 4 watts FM. There are 22 channels where 0.5 watt AM/SSB is allowed. There is also an 80-channel band in the 933-934 MHz region. Same goes for **Luxemburg** and **Liechtenstein.**

Turkey: Has the standard 40 channels, 4 watts AM.

Yugoslavia (former): No official regulations, but the standard 40 channels are in use, 4 watt transceivers.

In other areas of the world not specified here, CB can generally be assumed to be active on all of the 40 standard 27 MHz CB channels, or at least from 26.965 to 27.225 MHz (Channels 1 to 22). In some nations (such as Malta), CB is strictly forbidden-- however most nations do permit CB in one form or another. The only nations that have traditionally opposed CB are those with active revolutionary atmospheres. In such nations, the governments don't want the radios to be put to use for anti-government or underground broadcasting or communications purposes.

One last thing to remember is that just because there are regulations put out by governments, it doesn't mean that they are necessarily followed to the letter. Depending upon the nation and its interest in enforcement, CB operations may be quite distant from the actual rules.

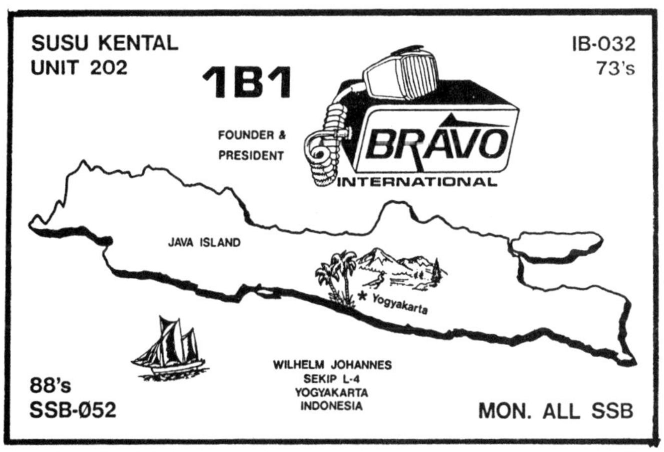

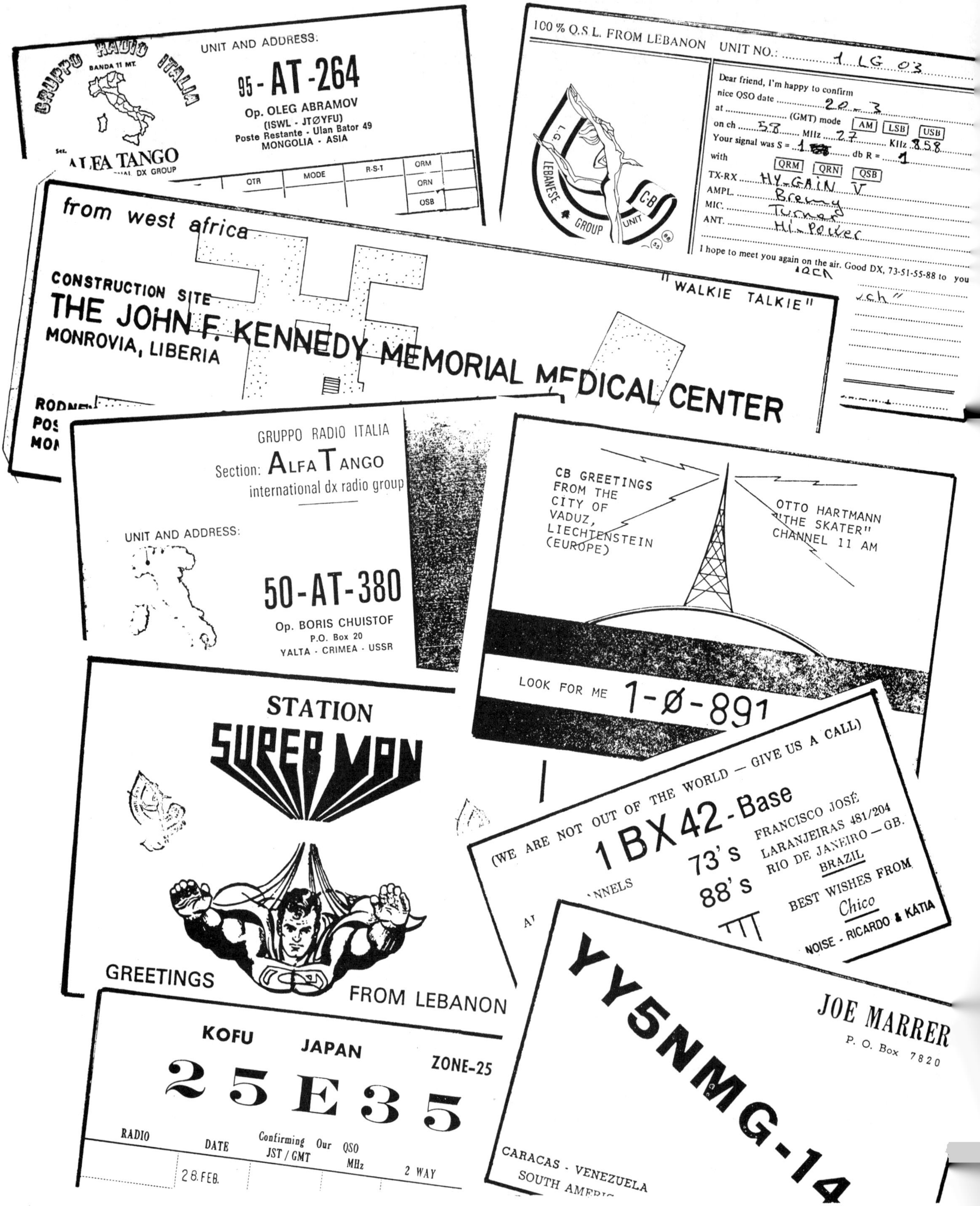
GRUPPO RADIO ITALIA
BANDA 11 MT.
UNIT AND ADDRESS:
95-AT-264
Op. OLEG ABRAMOV
(ISWL - JTØYFU)
Poste Restante - Ulan Bator 49
MONGOLIA - ASIA
ALFA TANGO
OTR
MODE
R-S-T
QRM
QRN
QSB
100 % Q.S.L. FROM LEBANON UNIT NO.: 1 LG 03
Dear friend, I'm happy to confirm
nice QSO date 20 3
at (GMT) mode AM LSB USB
on ch 58 MHz 27 KHz 858
Your signal was S = db R = 1
with QRM QRN QSB
TX-RX HY-GAIN V
AMPL.
MIC.
ANT.
I hope to meet you again on the air. Good DX, 73-51-55-88 to you
L.G. LEBANESE GROUP UNIT C.B
from west africa
CONSTRUCTION SITE
THE JOHN F. KENNEDY MEMORIAL MEDICAL CENTER
MONROVIA, LIBERIA
"WALKIE TALKIE"
GRUPPO RADIO ITALIA
Section: ALFA TANGO
international dx radio group
UNIT AND ADDRESS:
50-AT-380
Op. BORIS CHUISTOF
P.O. Box 20
YALTA - CRIMEA - USSR
CB GREETINGS FROM THE CITY OF VADUZ, LIECHTENSTEIN (EUROPE)
OTTO HARTMANN "THE SKATER" CHANNEL 11 AM
LOOK FOR ME 1-Ø-891
STATION
SUPER MAN
GREETINGS
FROM LEBANON
(WE ARE NOT OUT OF THE WORLD — GIVE US A CALL)
1 BX 42 - Base
73's
88's
FRANCISCO JOSÉ
LARANJEIRAS 481/204
RIO DE JANEIRO — GB.
BRAZIL
BEST WISHES FROM
Chico
NOISE - RICARDO & KATIA
KOFU
JAPAN
ZONE-25
25E35
RADIO
DATE
Confirming Our QSO
JST / GMT
MHz
2 WAY
28 FEB.
YY5NMG-14
JOE MARRER
P. O. Box 7820
CARACAS - VENEZUELA

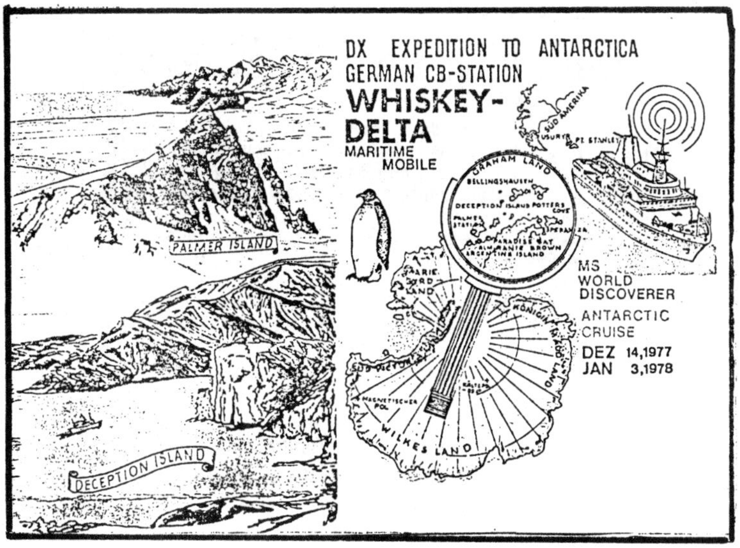

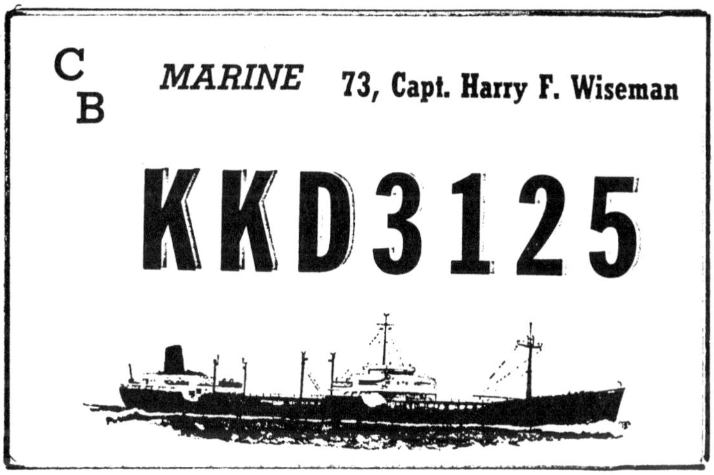

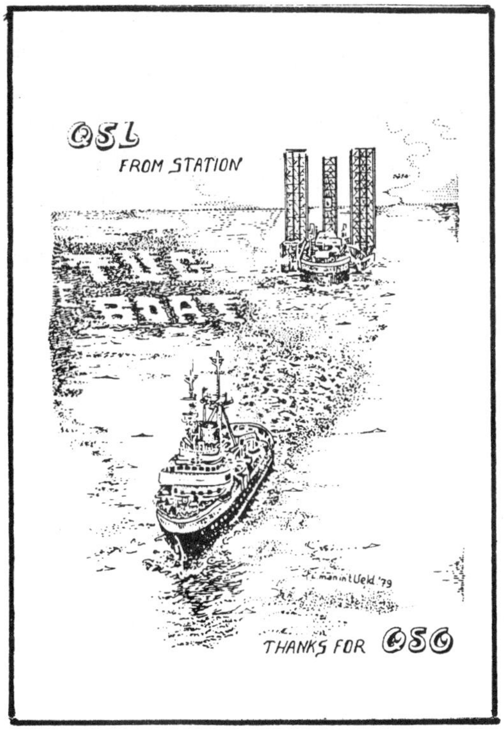

A bevy of highly prized QSL's from all the corners of the world are shown on these two pages. The ones above are from stations on ships in international waters.

Shooting Skip

Those who participate in the unauthorized practice of working DX (also known as shooting a little skip) on 27 MHz claim that it's an exciting challenge. Despite strong FCC restrictions against CB stations communicating past their immediate groundwave coverage, it's been going on since the early days of the CB service. I don't suggest that any operator violate the FCC's regulations, however it is nonetheless interesting to note the techniques used by the army of those who regularly visit Skipland.

They spend a lot of time just listening-- keeping their ears open, brain engaged, jaws in neutral gear. By listening half as much as they speak, they say they can get a clear picture of which stations are coming through in order to aim the beam and prepare properly. Skip experts realize that overseas stations are probably sitting there with a huge pileup of American and Canadian stations calling and attempting to attract their attention. Most likely the station will respond to the first station he can extract from the jumble-- so the idea seems to be to call him by announcing his ID once followed by the calling station's ID and location twice (Example: **Skyline** this is **Palm Tree** in Florida, **Palm Tree** in Florida.) Droning on is unnecessary as well as counterproductive. If a station is called three or four times like this doesn't respond, most operators go fishing for another station. There's little point in calling a station that has negative copy.

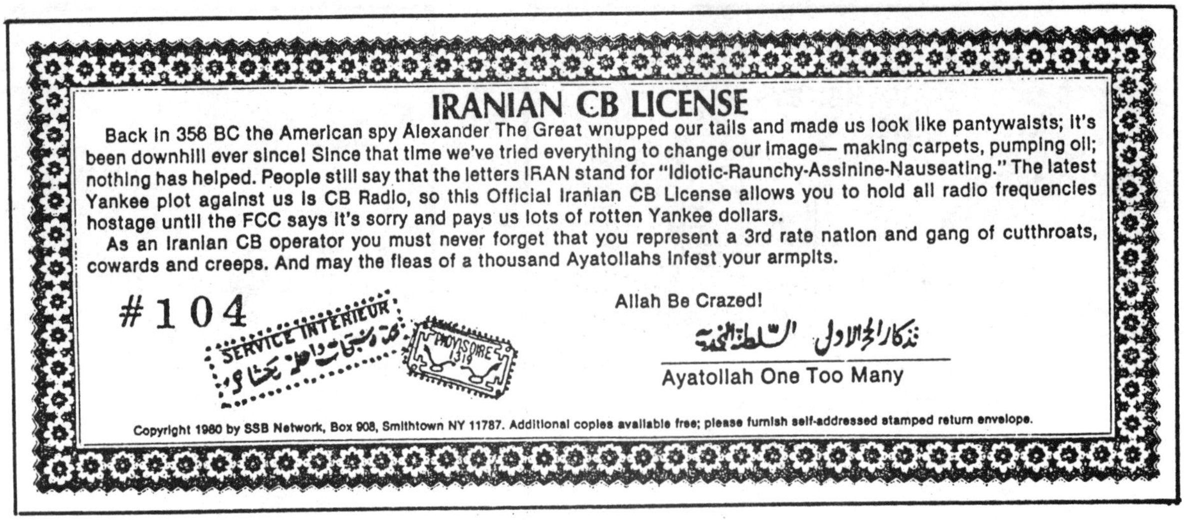
IRANIAN CB LICENSE

Back in 356 BC the American spy Alexander The Great wnupped our tails and made us look like pantywaists; it's been downhill ever since! Since that time we've tried everything to change our image— making carpets, pumping oil; nothing has helped. People still say that the letters IRAN stand for "Idiotic-Raunchy-Assinine-Nauseating." The latest Yankee plot against us is CB Radio, so this Official Iranian CB License allows you to hold all radio frequencies hostage until the FCC says it's sorry and pays us lots of rotten Yankee dollars.

As an Iranian CB operator you must never forget that you represent a 3rd rate nation and gang of cutthroats, cowards and creeps. And may the fleas of a thousand Ayatollahs infest your armpits.

#104

Allah Be Crazed!

Ayatollah One Too Many

Copyright 1980 by SSB Network, Box 908, Smithtown NY 11787. Additional copies available free; please furnish self-addressed stamped return envelope.

This gag "Iranian CB License" is the creation of the SSB Network.

A technique called "tail ending" is often used. This means that when the DX station is winding up his contact and turns it over for the other guy's goodbyes, the next station trying to work the DX station jumps right in ahead of the pack and gives a shout. It's not really horning in on anything since the contact was over anyway, but it does get a foot in the door before a zillion other stations start screaming for him. There's no point at all in calling the DX station while he is transmitting, although many do it anyway.

When contact is finally established, it's usually kept short and sweet since many others may be waiting. If it's a station from a place where English isn't the native tongue, DXperts avoid cutsie localized CB jargon and phrases that could be confusing. They save the "golleeee" and "mercy, mercy," for the locals that come through when Skipland is closed down. Discussions or opinions concerning politics are strictly avoided.

An overriding consideration with all who are successful at this is a sense of good sportsmanship and cooperation. It is, after all, a hobby-- not a competition or contest to the death. Supposedly, those who engage in it are enjoying themselves. Yet, some guys have the blood vessels standing out on their temples within ten minutes after they sit down at the rig-- not a very good way of relaxing. No matter how juicy a DX station may appear to be, it isn't worth having a conniption in order to snag him, or freaking out altogether if he gets away. He'll be back some other day.

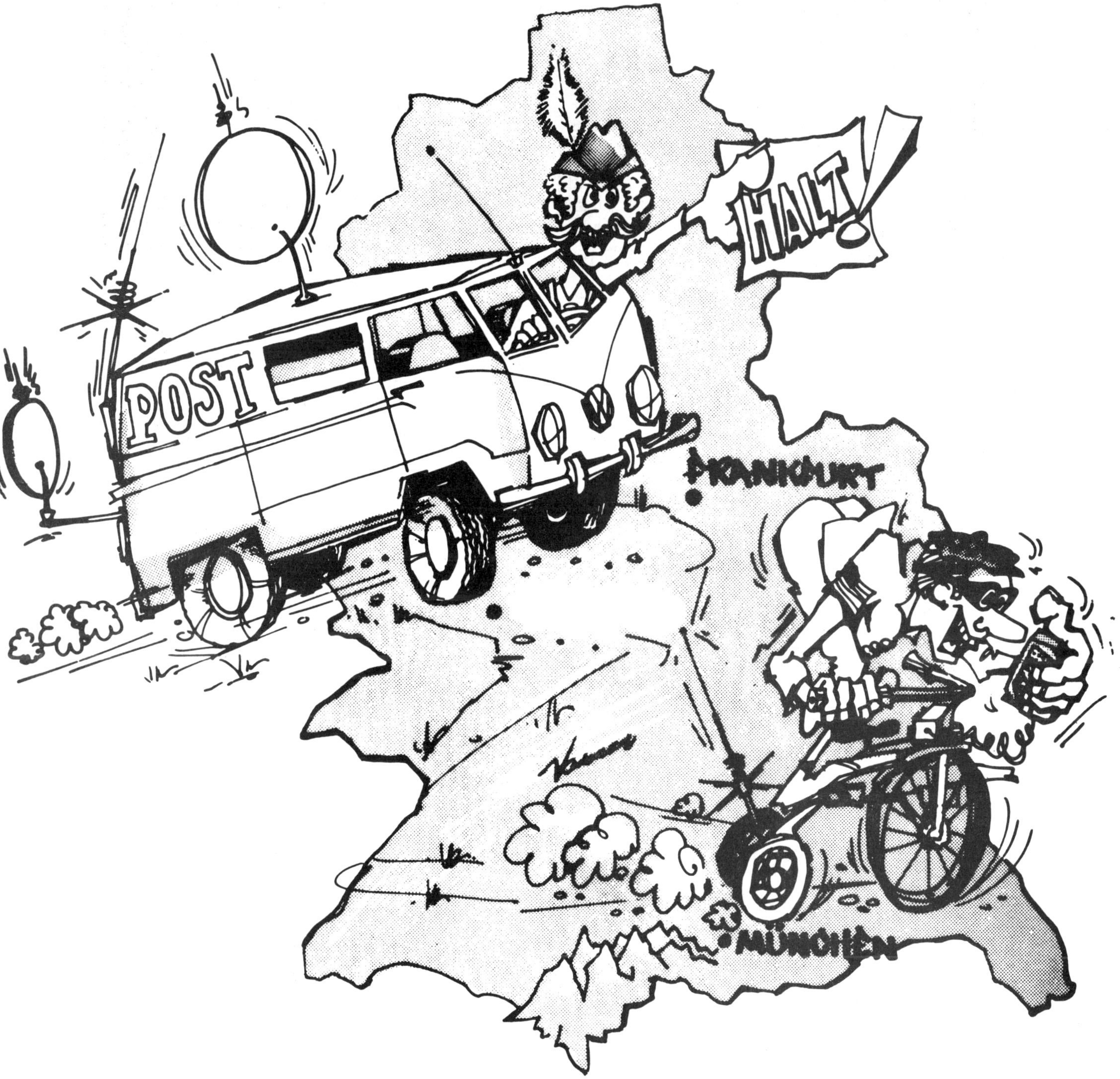

The West German authorities, constantly chasing down American GI's operating their CB's on unauthorized frequencies, inspired this zany illustration by Dave Varner.

Tomcat's CB Skipland Map

FCC Part 95 "Prohibited Uses" (Section 95.415(9)): "You must not use a CB station to communicate with, or attempt to communicate with any CB station more than 250 kilometers (155.3 miles) away." However, as you can see from this map of CB Skipland, it's really only 150 miles from coast-to-coast. Other maps are grossly inaccurate.

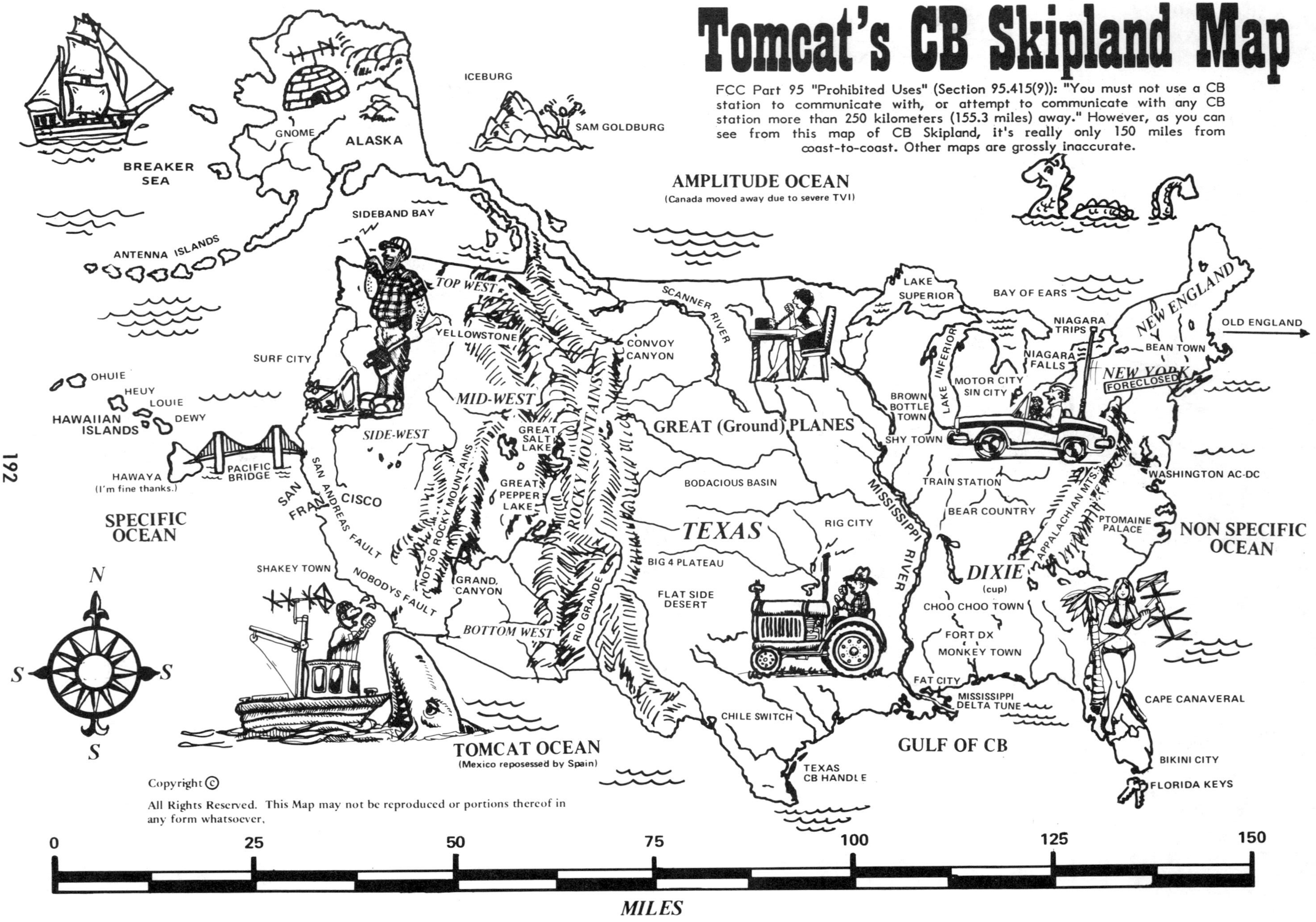

International Time

Communications logging is usually done in terms of Coordinated Universal Time (UTC), also known as Greenwich Mean Time (GMT). For logging purposes, the date changes when the time becomes 0000 UTC in your local area. For instance, in the Eastern Standard Time Zone at 6 p.m. on 28 January, a station would be logged at 2300 UTC on 28 January. But a station logged two hours later at 8 p.m., would be entered in your records as 0100 UTC on **29** January. The following chart converts UTC/GMT to local time zones:

24-hour Greenwich Mean Time (GMT)	Eastern Standard Time (EST)	Eastern Daylight Time (EST)	Central Standard Time (CST)	Central Daylight Time (CDT)	Mountain Standard Time (MST)	Mountain Daylight Time (MDT)	Pacific Standard Time (PST)	Pacific Daylight Time (PDT)
0000	7 P.M.	8 P.M.	6 P.M.	7 P.M.	5 P.M.	6 P.M.	4 P.M.	5 P.M.
0100	8 P.M.	9 P.M.	7 P.M.	8 P.M.	6 P.M.	7 P.M.	5 P.M.	6 P.M.
0200	9 P.M.	10 P.M.	8 P.M.	9 P.M.	7 P.M.	8 P.M.	6 P.M.	7 P.M.
0300	10 P.M.	11 P.M.	9 P.M.	10 P.M.	8 P.M.	9 P.M.	7 P.M.	8 P.M.
0400	11 P.M.	12 Mid.	10 P.M.	11 P.M.	9 P.M.	10 P.M.	8 P.M.	9 P.M.
0500	12 Mid.	1 A.M.	11 P.M.	12 Mid.	10 P.M.	11 P.M.	9 P.M.	10 P.M.
0600	1 A.M.	2 A.M.	12 Mid.	1 A.M.	11 P.M.	12 Mid.	10 P.M.	11 P.M.
0700	2 A.M.	3 A.M.	1 A.M.	2 A.M.	12 Mid.	1 A.M.	11 P.M.	12 Mid.
0800	3 A.M.	4 A.M.	2 A.M.	3 A.M.	1 A.M.	2 A.M.	12 Mid.	1 A.M.
0900	4 A.M.	5 A.M.	3 A.M.	4 A.M.	2 A.M.	3 A.M.	1 A.M.	2 A.M.
1000	5 A.M.	6 A.M.	4 A.M.	5 A.M.	3 A.M.	4 A.M.	2 A.M.	3 A.M.
1100	6 A.M.	7 A.M.	5 A.M.	6 A.M.	4 A.M.	5 A.M.	3 A.M.	4 A.M.
1200	7 A.M.	8 A.M.	6 A.M.	7 A.M.	5 A.M.	6 A.M.	4 A.M.	5 A.M.
1300	8 A.M.	9 A.M.	7 A.M.	8 A.M.	6 A.M.	7 A.M.	5 A.M.	6 A.M.
1400	9 A.M.	10 A.M.	8 A.M.	9 A.M.	7 A.M.	8 A.M.	6 A.M.	7 A.M.
1500	10 A.M.	11 A.M.	9 A.M.	10 A.M.	8 A.M.	9 A.M.	7 A.M.	8 A.M.
1600	11 A.M.	12 Noon	10 A.M.	11 A.M.	9 A.M.	10 A.M.	8 A.M.	9 A.M.
1700	12 Noon	1 P.M.	11 A.M.	12 Noon	10 A.M.	11 A.M.	9 A.M.	10 A.M.
1800	1 P.M.	2 P.M.	12 Noon	1 P.M.	11 A.M.	12 Noon	10 A.M.	11 A.M.
1900	2 P.M.	3 P.M.	1 P.M.	2 P.M.	12 Noon	1 P.M.	11 A.M.	12 Noon
2000	3 P.M.	4 P.M.	2 P.M.	3 P.M.	1 P.M.	2 P.M.	12 Noon	1 P.M.
2100	4 P.M.	5 P.M.	3 P.M.	4 P.M.	2 P.M.	3 P.M.	1 P.M.	2 P.M.
2200	5 P.M.	6 P.M.	4 P.M.	5 P.M.	3 P.M.	4 P.M.	2 P.M.	3 P.M.
2300	6 P.M.	7 P.M.	5 P.M.	6 P.M.	4 P.M.	5 P.M.	3 P.M.	4 P.M.

International Phonetic Alphabet

When passing communications traffic, it is often desirable to spell out words and names. Operators attempting to do this sometimes find that it's necessary to spell out words using phonetics so that the person receiving the information can get the spelling correctly. A standard set of international phonetics was therefore established (originally for use by military and aviation interests) that is especially suited to this purpose. When there's a need to spell something with phonetics-- or give sideband ID numbers through rough QRM, these should be of help:

Alfa
Bravo
Charlie
Delta
Echo
Foxtrot
Golf
Hotel
India
Juliet
Kilo
Lima
Mike
November
Oscar
Papa
Quebec
Romeo
Sierra
Tango
Uniform
Victor
Whiskey
X-ray
Yankee
Zulu

Using these phonetics, for example, the name "Jones" would be spelled out as "Juliet-Oscar-November-Echo-Sierra." Sideband numbers would be stated along the lines of "Sierra-Sierra-Bravo One-Three."

Numerical Sideband Prefixes

Members of various Sidebanding groups sometimes add numerical prefixes to their sideband ID numbers, especially when working skip. This gives the distant station a quick idea of the station's general location. Therefore, a station in California with sideband numbers "GXN-288" might also at times identify as "31GXN-288." The following prefixes have been used often:

1 Delaware
2 Pennsylvania
3 New Jersey
4 Georgia
5 Connecticut
6 Massachusetts
7 Maryland
8 South Carolina
9 New Hampshire
10 Virginia & D.C.
11 New York
12 North Carolina
13 Rhode Island
14 Vermont
15 Kentucky
16 Tennessee
17 Ohio
18 Louisiana
19 Indiana
20 Mississippi
21 Illinois
22 Alabama
23 Maine
24 Missouri
25 Arkansas
26 Michigan
27 Florida
28 Texas
29 Iowa
30 Wisconsin
31 California
32 Minnesota
33 Oregon
34 Kansas
35 West Virginia
36 Nevada
37 Nebraska
38 Colorado
39 North Dakota
40 South Dakota
41 Montana
42 Washington
43 Idaho
44 Wyoming
45 Utah
46 Oklahoma
47 New Mexico
48 Arizona
49 Alaska
50 Hawaii
51 Canadian Maritimes
52 Quebec
53 Greenland
54 Ontario
55 Manitoba
56 Saskatchewan
57 Alberta
58 British Columbia
59 Mexico
60 West Indies
61 Puerto Rico
62 Panama/Central Amer.
63 Venezuela
64 Brazil
65 Argentina
66 Chile
67 Australia
68 Guam
69 Japan
70 England
71 West Germany
72 France
73 Italy
74 Sweden
75 Ireland
76 Spain
78 Bermuda

16

Troubleshooting Basics

Most CB repairs are best left for being attended to by qualified service technicians. Still, not everything that goes wrong with a CB rig is beyond the realm of being straightened-out at home. Like the CB'er who took his rig to the service technician to complain that something was wrong. After a couple of days of anxious waiting, it was returned with a note saying that there wasn't anything the matter that switching the PA/CB switch to CB wouldn't cure. A service technician once confided in me that many rigs brought in for repair could have been taken care of at home, often in a few minutes by a person with no technical background.

Therefore, lets go through some of the more commonly encountered problems that you might try to tackle at home before you decide to let a professional tackle the job. Even if you find that it's still something you can't or don't want to fix, you'll possibly at least get a better idea as to what the problem might be. Naturally, any attempts on your part to poke around the innards of a CB rig should be done only with the power cord removed from the wall socket so you aren't rapidly dispatched to that Big Coffee Break In The Sky. Even the 12 volts in your mobile unit can be rather nasty, and should be disconnected if you're going to tinker with the mobile transceiver. Work only with dry hands around any electrical or electronic equipment.

Tomcat's Sure-Fire
Olde Tyme Hoodoo Cure

for annoying inter-ference from neighbor's CB sets

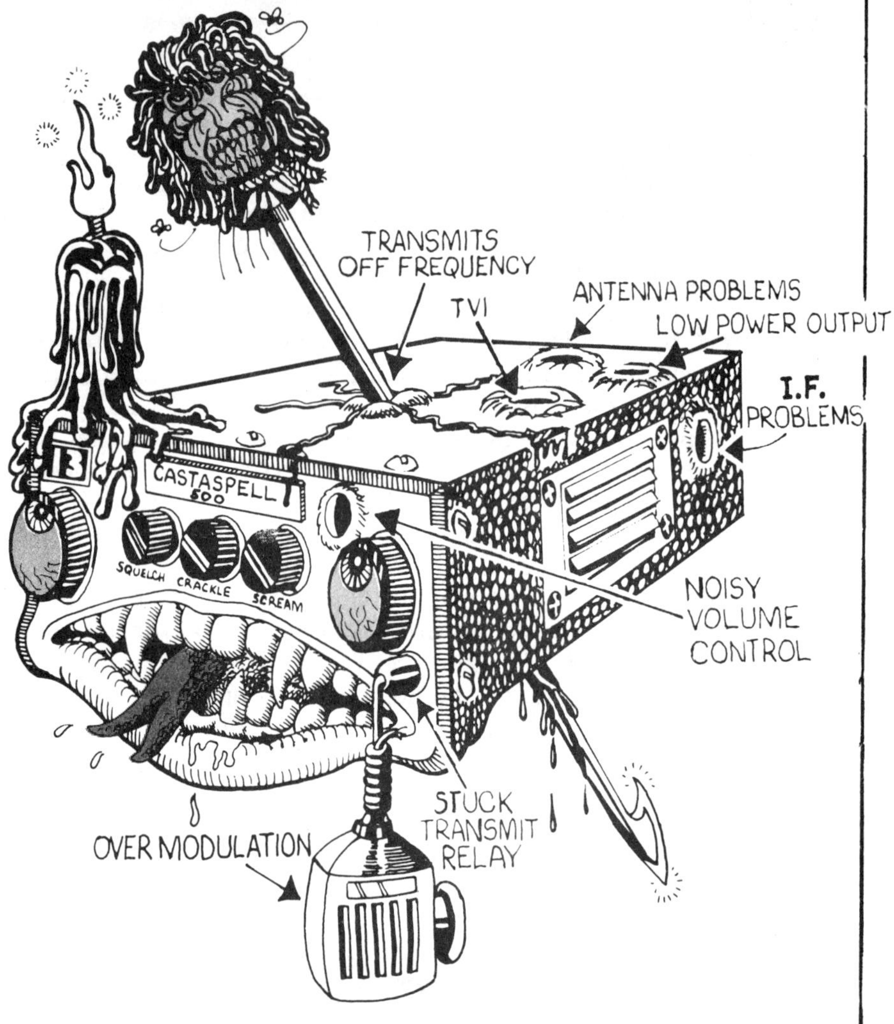

Sometimes one of the problems isn't that your set isn't broken, but the guy next door has one that's working too well. Every time your neighbor comes on the air, he wipes you right out. Troubleshooting your own rig is meaningless-- but you might try "fixing" his set so it doesn't work so well. Supposedly this old trick has been passed down through countless generations of CB'ers. I got it at small personal risk while vacationing in the alligator infested bayous of the Hudson River. Old **Mama Mojo** came creaking out of her hut, winked, and handed me these instructions on how to get those loud and pesky CB neighbors off the band-- like the guy down the block who spends all day asking for 10-36's and "breaks," while jamming 15 channels on your rig with each shout.

Mama Mojo says to hoodoo him off the air! Get a cardboard box and make an effigy of his rig. Then get some long hatpins (the kind with the big knob at the end). **Mama Mojo** says to stick them through various parts of the cardboard effigy rig to produce a variety of horrible mishaps in the other guys's rig-- overmodulation, low power output, TVI, off frequency operation, etc. **Mama Mojo** even gave me the secret pin-insertion locations for best results.

Might help if you danced around a bit while doing this, maybe chanting something fierce as you thrust the pins into the set. One variation on this was passed along by Dr. Wun Hung Low, noted CB'ing acupuncturist. He says that if your own rig breaks, you can do the same thing with long golden pins and maybe it will cure the problem!

The advice given here is not intended to replace the service of a qualified service technician-- I'm only trying to save you a trip there if there's a minor and easily correctable superficial problem. CB rigs do, however, develop serious internal disasters, and if nothing here helps then you'd best let an expert have a go at the problem. It's really not a good idea to twiddle around with the innards of a CB rig (even with the power disconnected) because you could cause more problems than you started out with while also voiding the factory warranty.

If you check over your gear as indicated here, you won't cause any damage to yourself or the CB set-- and you could save some greenstamps.

1. Set is absolutely dead, won't even gurgle, nothing lights up, it's truly and completely kaput. Maybe you just thought you turned the power on, but really didn't. Is the fuse OK? If not, replace it with a spare of the same size/rating (hope you keep some on hand). Check base/mobile power connections. In mobile installations, check the ground connection. Check power cord (while it's unplugged) for broken or frayed wires, damaged insulation, bent pins on plug. Check household circuit breaker on the line you're use to power the rig. Remember, in some mobile installations the rig works only when the ignition is "on."

2. Fuse is blown. Do you have the correct size/rated fuse in the set? Maybe

the polarity of your mobile hookup is reversed. Could be something shorting out inside the unit, but maybe the power cord is fraying and zapping everything-- check it (unplugged from the wall) where it enters the rig. Maybe screws from the mobile mounting bracket are shorting to the innards of the rig.

3. CB keeps winking and blinking at you, its lights keep going on and off as the receiver goes on/off in unison. Sounds like a loose power connection. Check out the power source, fuse contacts, power plug as it enters the CB transceiver.

4. Rig is all lit up, but isn't making any sounds. Mike connector may not be inserted correctly, or have developed a wiring problem-- could require a new mike connector. CB/PA switch in wrong position. Squelch, audio gain (volume control), or RF gain control improperly adjusted. Headphones inserted in jack.

5. Only stations received are those within close proximity, set seems to be operating with some sort of internal squelch that quiets it between incoming transmissions although set's squelch is "open." Audio may be distorted. This can be the result of a high SWR-- that is to say that something in your antenna system is out of whack. Check antenna connector, cable, the antenna itself. A wet or iced antenna could temporarily raise the SWR. See if RF gain control is properly adjusted. Voltage may be weak. Unit might need realignment by a service technician.

6. Set is on, but it seems locked into transmit mode and can't be made to receive. Could be a problem in the mike's push-to-talk (PTT) button or the mike connector. Could also be an internal problem needing a technician. I once had a mobile rig that pulled this stunt as soon as I turned on the set in cold weather. After ten minutes of driving with the car's heater on, it worked properly. Turned out to be a temperature sensitive relay in the rig-- a tech would be needed to fix this.

7. Set is on but incoming signals are garbled and can't be understood. If you've got an AM rig, youve probably tuned to a sideband frequency. Switch over Channel 19 or another AM channel and see if things improve. Could also be a torn loudspeaker cone, or an audio circuit problem, or set out of alignment.
8. Set is on but all incoming signals are buried 'neath a bad howl. If rig is an AM/SSB type, switch mode selector knob to LSB or USB.
9. On SSB, stations won't clarify. If you're in LSB mode, the station you're trying to clarify is probably using USB-- or else is using a slider and has shifted too far off center slot for clarification.
10. Other stations complain that there's a squeal on your signal. Might be RF feedback caused by high SWR. Check antenna system as in #5.
11. Set seems operational, but appears as though all of a sudden everybody you normally hear is off the air. Check the channel selector, you may be on the wrong channel.
12. About all the set will do is croak out an irritating buzz. If you're trying to operate a mobile rig at home from a power supply, then you're probably not using one that delivers filtered DC. Or, remove antenna connector and replace it with a new one. Could be severe power line noise, or from a nearby fluorescent or neon lighting fixture-- or even a defective light bulb.
13. The S-meter indicates that there are incoming signals, but there's no audio emerging from the loudspeaker. May be a faulty 'speaker, or the connections to the speaker. Try headphones or an external speaker. Could also be misadjusted, dirty, or defective volume control. Check adjustment of squelch control. Might well be an internal circuitry problem that needs professional attention on a service bench.

14. Smoke seems to be drifting from the rig, and you can smell something burning. For starters, switch off the set immediately and disconnect it from its power source. Maybe the fuse was wrong (too high a rating) for the set-- the right one might have blown before you began blowin' smoke. Are the power leads connected with the correct polarities? If everything looks OK, better let a technician look it over before you attempt to turn it on again.

15. Set makes an eerie humming noise, but doesn't do much else. If you've been running the rig from house current, try it from the 12 VDC in the car. If that cures the problem, then its something inside the rig for a pro to work on. Maybe the mike isn't connected properly. If it is, then let a technician work it over.

16. Works very well at one end of the band but operation definitely poorer at the opposite end. Providing the antenna hasn't been damaged (check it), chances are that the antenna is either slightly too short or too long. Using an SWR bridge, and with the rig in AM mode, tune the antenna either for the Channel 19, or else for your most-used channel.

17. Receiver is OK, but other operators complain that you sound "scratchy" or your audio is cutting on/off. Probably something fouled up in the mike connector-- bad connection or wires shorting.

18. An AM rig with no transmitter modulation although there's a carrier on the air. May be a bad mike button or mike connection. Mike gain control not properly adjusted.

19. An SSB rig that receives OK but doesn't seem to want to transmit. Possibly the mike gain is turned down. Switch over to AM mode to confirm that there's a carrier and the whole transmitter hasn't gone to Sheboygan.

20. Other operators complain that you're "splashing" them on channels adjacent to the one on which you're operating. Probably you're overmodulating, back down on the rig's audio gain-- or your power mike's gain.

17

In-Transit Mobile Channels

Mobile operating in your own local area may be as easy for you as wandering from the TV room to the fridge, but when you aim that front bumper into distant areas you'll soon realize that things may not be the same-- that illogical operators in other areas had the unmitigated gall and poor judgment to use channels that are 90° out of phase with all you know to be good, wholesome, right, moral, ethical, honest, and rational.

Yes, you're aware that none of these strange customs make any sense at all-- and far be it from me to try to explain how or why such customs were ever started, or give you the name of the crazed person who institute such radical policies. I mean, what's the matter with good 'ol Channel 19 as a mobile in-transit channel? Why doesn't everybody just pack up and go back there instead of hanging out on other channels? One person suggested that Channel 19 is where the tourists congregate-- the other channels are used by the local residents laughing at the funny looking out-of-towners.

Now that we agree that it's absurd, let's face the fact that in order to get the maximum use from your CB rig while you're on the road, you're going to have to accept the idea that you may have to do things their way. That's what this part of the book is about; to clue you in.

While Channel 19 is generally a channel that you can try in the blind without worrying about making too big a fool out of yourself no matter where you are, there are alternate mobile channels in use instead of or in addition to Channel 19. In some areas, there are so many truckers on Channel 19 that passenger vehicles have fled to other channels in order to get a word in edgewise. Sometimes, several different highways run parallel to one another at close range and while one may use Channel 19, the others utilize different

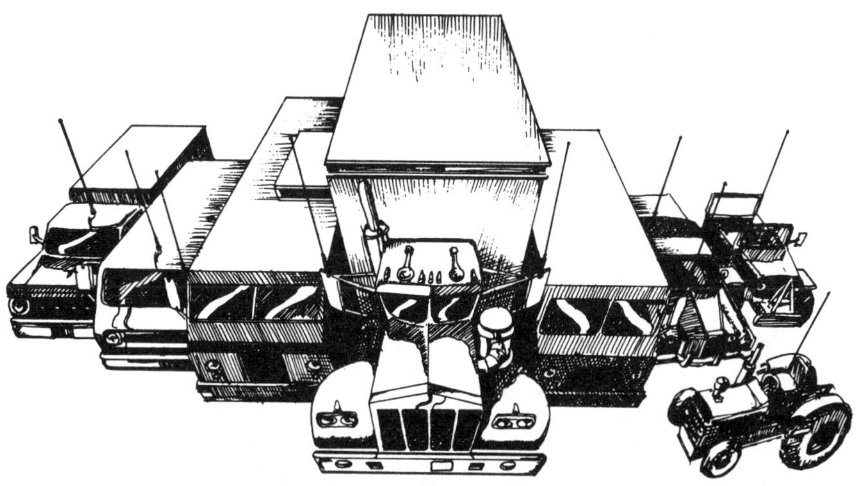

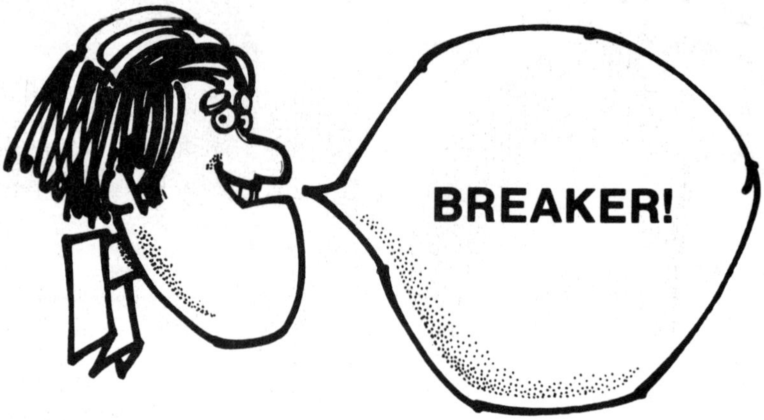

channels just to keep everything sorted-out. Very often, Channel 10 is selected to play second fiddle to Channel 19.

Some roads in our directory indicate several channels in use. This may be accounted for because: 1) the truckers are on Channel 19 while the 4-wheelers are elsewhere, or 2) channels may change over different sections of the road-- when in doubt, try those listed or inquire on Channels 9 or 19 for for further information.

The listings here are based upon my own observations plus those sent in by correspondents which I assume to be correct and current. They relate to AM (as opposed to SSB) communications. A listing that reads "**8**/19" means that on Route/Highway/Interstate 8, Channel 19 is the frequency to monitor for traffic conditions, Smokey Breaks, and chit-chat between drivers. A listing reading "**8**/6-10-19" would indicate that communications for the road appear to exist on Channels 6, 10, and 19. On roads for which no listing is given, communications are presumed to be on Channel 19. Readers are invited to furnish corrections/updates/additional roads/channels to this directory:

In some areas, truckers are so active on Channel 19 that other drivers have fled to alternate channels. Still, Channel 19 is the best place to hear about what "Smokey" is doing that day.

Folks driving RV's have long liked to look for one another on Channel 13. It's strictly unofficial, but fun!

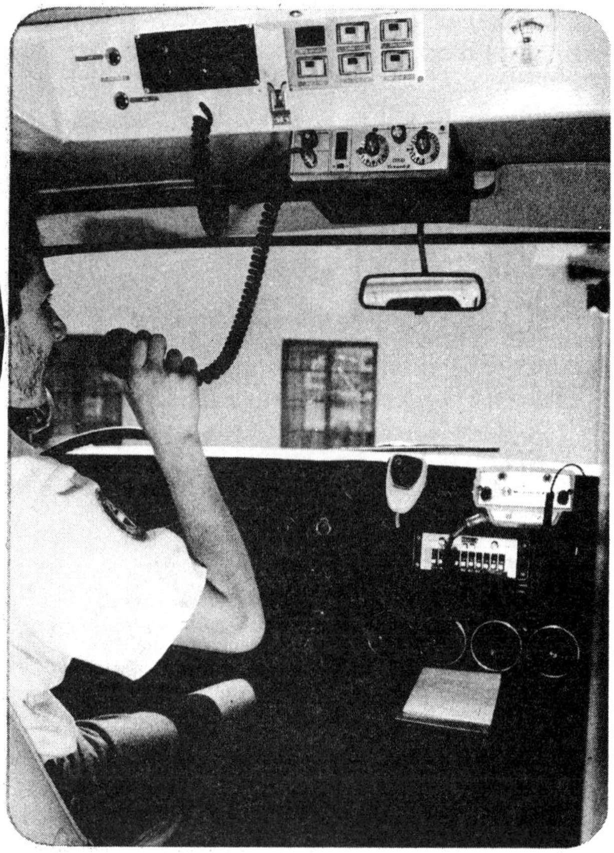

Alabama: All roads seem to use Channels 10-11-19-21.
Alaska: All roads use Channels 19-21.
Arizona: **8**/14-19-21; **10**/19-21; **17**/21; **40**/17-19-21.
Arkansas: **30**/10-11-19; **40**/10-19; **55**/10-19.
California: **5**/5-6-10-15-17-21; **8**/19-21; **10**/19-21; **15**/17-19-21; **17**/21; **40**/5-19-21; **80**/17-19-21; **101**/17-18-19; **395**/19-21.
Colorado: **25**/19-21; **70**/10-19-21; **80S**/19; **287**/19-21.
Connecticut: All roads Channels 10-19; **95**/19; **Merritt Pkwy**/15; **Wilbur Cross Pkwy**/15.
Delaware: All roads Channel 19.
Distr. of Columbia: All roads Channel 19; **Canal Rd.**/4; **McArthur Rd**/4.
Florida: **10**/19; **11**/15; **75**/10-11-15; **95**/10-11-19.
Georgia: **19**/10-11; **20**/10-11-19; **75**/10-11-15-19; **95**/10-11-19.
Hawaii: All roads Channel 19.
Idaho: **15**/11-19; **80N**/19-20-21; **90**/17-19.
Illinois: All roads Channels 10-19-21.
Indiana: All roads Channels 10-19-21; **64**/12-19.
Iowa: **29**/19; **35**/10-19; **69**/10-11-19; **80**/10-13-19.
Kansas: **35**/3-7-19-20; **50**/20; **70**/5-10-21.
Kentucky: All roads Channels 10-19; **75**/12-19.
Louisiana: All roads Channels 10-19; **20**/11-17.
Maine: All roads Channels 10-19.
Maryland: All roads Channel 19.
Massachusetts: All roads Channel 19.
Michigan: All roads Channels 10-19.
Minnesota: **35**/10; **69**/10; **90**/10-19; **94**/10-19.
Mississippi: All roads Channels 10-19; **20**/11.
Missouri: **14**/15; **29**/10-14-15-19; **44**/10-19; **55**/10-19; **66**/10-19; **70**/10-19-20

Montana: All roads Channel 19; **94**/11-15-19.
Nebraska: All roads Channels 10-19-20; **17**/19-20-21.
Nevada: **15**/17-19-21; **80**/3-11-17-19-21.
New Hampshire: All roads Channels 10-19.
New Jersey: All roads Channel 19; **Palisades Pkwy**/15.
New Mexico: **10**/17-19-21; **25**/19-21; **40**/19-21; **54**/15-17-19.
New York: **Hutchinson Pkwy.**/15; **Palisades Pkwy.**/15; **Sprain Pkwy.**15; **Taconic Pkwy.**/23. N.Y. City: **W. Side Dr.**/15-19; **Bronx River Pkwy.**/15; **FDR Drive**/19; **Belt Pkwy**/10; **Cross Isl. Pkwy.**/10; **Gr. Centr. Pkwy.**/12-19; **Interboro Pkwy.**/12; Long Island: **N. State Pkwy.**/12; **S. State Pkwy.**/10; **Sunrise Hwy.**/10; **Sagtikos Pkwy.**/10; **Wantagh Pkwy.**/10; **25**/12; **25A**/10; **27 & 27A**/10; **135**/10; **347**/12; **454**/10-12; **495**/19.
North Carolina: **40**/10-11-19; **85**/10-11-12-19; **95**/10-11-19; **421**/10-11-12-19.
North Dakota: All roads Channels 11-19.
Ohio: All roads Channel 19; **70 & 75**/15-21.
Oklahoma: All roads Channels 15-19; **35**/10-15-19-20-21; **40**/10-15-19-21.
Oregon: **5**/6-17-19-21; **80N**/11-17-19.
Pennsylvania: All roads Channel 19; **22**/5-19; **30**/5-15-19.
Puerto Rico: All roads Channel 19.
Rhode Island: All roads Channels 10-19.
South Carolina: All roads Channels 10-19; **20**/10-11-19; **26**/10-19; **85**/10-11-19; **95**/10-11-19.
South Dakota: All roads Channel 19.
Tennessee: All roads Channels 11-19; **24**/10-11; **40**/10-11-19; **65**/10-11; **75**/10-11; **78**/10.
Texas: All roads Channels 19-21; **10**/17-19-21; **20**/19; **24**/10-11-19; **30**/15-19; **35**/15-19; **40**/10-11-19; **45**/15-19; **65**/10-11-19; **75**/11-19; **78**/10-11.
Utah: All roads Channels 10-17-19-21.
Vermont: All roads Channels 10-19.
Virginia: All roads Channel 19; **81**/11-12-19.
Virgin Islands: All roads Channel 19.
Washington: 2E/2-10-14-19; **2W**/3-14-15; **5**/17-19; **10**/17-19; **12**/17-19; **90**/17-19. (Logging trucks- All roads Channel 14.)
West Virginia: All roads Channel 19.
Wisconsin: All roads Channel 19; **94**/11-19.
Wyoming: All roads Channel 19; **25**/19; **80**/19-20-21; **90**/19.
All areas: Recreational vehicles on Channel 13.
Canada: All roads Channels 1-19.
Quebec (W. of Montreal) on Channel 10.
Maritime Provinces on Channel 10.
Mexico: All roads on Channel 10.

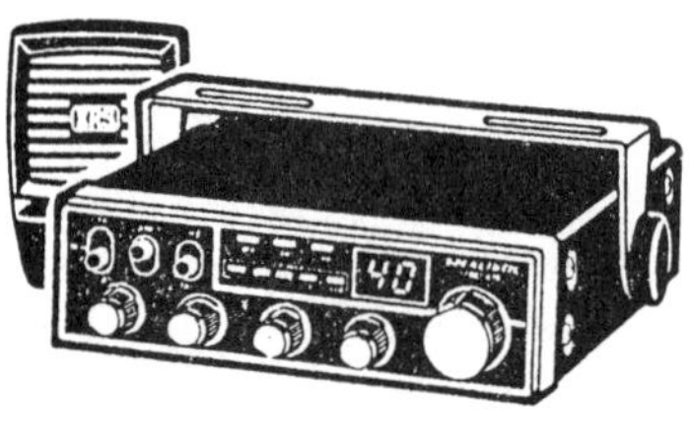

Radio Shack's Realistic TRC-474 mobile transceiver, a popular low-priced unit that offers a Channel 9 priority feature.

18

The FCC's CB Regulations

Probably no set of federal regulations has gone through so many different versions, modifications, and clarifications as the FCC's CB rules, which are presently known as Part 95 (previously known as Part 19). Not only are they in an almost constant state of flux, their enforcement has been very spotty and erratic, seeming to be based primarily upon budgetary limitations, manpower availability, and the whims of those in charge of such matters. Sometimes there are major enforcement campaigns, marked by the agency directing concentrated efforts in several metropolitan areas for a few months-- a few hapless rule violators and dealers in illegal equipment are singled out as examples for the news media to report. This may be followed by a lengthy period of relative quiet from the agency's minions. Still, going back to the late 1950's, when 27 MHz CB got its start, it's always been the radio service with the most rule violators-- probably because the general public (for whose benefit the CB service was said to be created) has insisted on CB radio reflecting its own wants and needs, seeing little sense in most of the FCC's rules which seem to have been prepared for use in a service the FCC would very much like to see in operation. This has forced the public to, in a very real sense, create its own radio service which, unfortunately, touches upon the FCC rules in only the very broadest sense. Presented here are the FCC's CB rules in the version that was in effect as we go to press. This is the so-called "Plain English Rules" version, as opposed to the coexisting more-formal rendition which, by self-definition, isn't written in plain English.

Plain English Rules — Citizens Band Radio Service

prepared by the **Federal Communications Commission,** Washington, D.C.

Part 95, Subpart D — Citizens Band (CB) Radio Service

General Provisions

§95.401 (CB Rule 1) What is the Citizens Band (CB) Radio Service?

The CB Radio Service is a private, two-way, short-distance voice communications service for personal or business activites. The CB Radio Service may also be used for voice paging.

§95.402 (CB Rule 2) How do I use these rules?

(a) You must comply with these rules (See CB Rule 21 §95.421, for the penalties for violations) when you operate a station in the CB Service from:

(1) Within or over the territorial limits of places where radio services are regulated by the FCC (see CB Rule 5, §95.405);

(2) Aboard any vessel or aircraft registered in the United States; *OR*

(3) Aboard any unregistered vessel or aircraft owned or operated by a United states citizen or company.

(b) Your CB station must comply with technical rules found in Subpart E of Part 95.

(c) Where the rules use the word "you", "you" means a person operating a CB station.

(d) Where the rules use the word "person", the rules are concerned with an individual, a corporation, a partnership, an association, a joint stock company, a trust, a state, territorial or local government unit, or other legal entity.

(e) Where the rules use the term "FCC", that means the Federal Communications Commission.

(f) Where the rules use the term "CB station", that means a radio station transmitting in the CB Radio Service.

§95.403 (CB Rule 3) Am I eligible to operate a CB station?

You are authorized to operate a CB station unless:

(a) You are a foreign government, a representative of a foreign government, or a federal governmental agency; *OR*

(b) The FCC has issued a cease and desist order to you, and the order is still in effect.

§95.404 (CB Rule 4) Do I need a license?

You do not need an individual license to operate a CB station. You are authorized by this rule to operate your CB station in accordance with the rules in this Subpart.

§95.405 (CB Rule 5) Where may I operate my CB station?

You are authorized to operate your CB station from:

(a) Within or over any area of the world where radio services are regulated by the FCC. Those areas are within the territorial limits of:

(1) The fifty United States.

(2) The District of Columbia.

Caribbean Insular areas

(3) Commonwealth of Puerto Rico.

(4) Navassa Island.

(5) United States Virgin Islands (50 islets and cays).

Pacific Insular areas

(6) American Samoa (seven islands).

(7) Baker Island.

(8) Commonwealth of Northern Mariana Islands.

(9) Guam Island.

(10) Howland Island.

(11) Jarvis Island.

(12) Johnston Island (Islets East, Johnston, North and Sand).

(13) Kingman Reef.

(14) Midway Island (Islets Eastern and Sand).

(15) Palmyra Island (more than 50 islets).

(16) Wake Island (Islets Peale, Wake and Wilkes).

(b) Any other area of the world, except within the the territorial limits of areas where radio services are regulated by —

(1) An agency of the United States other than the FCC. (You are subject to its rules.)

(2) Any foreign government. (You are subject to its rules.)

(c) An aircraft or ship, with the permission of the captain, within or over any area of the world where radio services are regulated by the FCC or upon or over international waters. You must operate your CB station according to any applicable treaty to which the United States is a party.

§95.406 (CB Rule 6) Are there any special restrictions on the location of my CB station?

(a) If your CB station is located on premises controlled by the Department of Defense you may be required to comply with additional regulations imposed by the commanding officer of the installation.

(b) If your CB station will be constructed on land of environmental or historical importance (such as a location significant in American history, architecture or culture), you may be required to provide information and to comply with §§1.1305 through 1.1319 of the FCC's Rules.

How To Operate a CB Station

§95.407 (CB Rule 7) On what channels may I operate?

(a) Your CB station may transmit only on the following channels (frequencies):

Channel	Frecuency (megahertz-MHz)
1	26.965
2	26.375
3	26.985
4	27.005
5	27.015
6	27.025
7	27.035
8	27.055
9	[1]27.065
10	27.075
11	27.085
12	27.105
13	27.115
14	27.125
15	27.135
16	27.155
17	27.165
18	27.175
19	27.185
20	27.205
21	27.215
22	27.225
23	27.255
24	27.235
25	27.245
26	27.265
27	27.275
28	27.285
29	27.295
30	27.305
31	27.315
32	27.325
33	27.335
34	27.345
35	27.355
36	27.365
37	27.375
38	27.385
39	27.395
40	27.405

[1]See paragraph (b) of this section.

(b) Channel 9 may be used only for emergency communications or for traveler assistance.

(c) You must, at all times and on all channels, give priority to emergency communication messages concerning the immediate safety of life or the immediate protection of property.

(d) You may use any channel for emergency communications or for traveler assistance.

(e) You must share each channel with other users.

(f) The FCC will not assign any channel for the private or exclusive use of any particular CB station or group of stations.

(g) The FCC will not assign any channel for the private or exclusive use of CB stations transmitting single sideband or AM.

§95.408 (CB Rule 8) How high may I put my antenna?

(a) "Antenna" means the radiating system (for transmitting, receiving or both) and the structure holding it up (tower, pole or mast). It also means everything else attached to the radiating system and the structure.

(b) If your antenna is mounted on a hand-held portable unit, none of the following limitations apply.

(c) If your antenna is installed at a fixed location, it (whether receiving, transmitting or both) must comply with either one of the following:

(1) The highest point must not be more than 6.10 meters (20 feet) higher than the highest point of the building or tree on which it is mounted; *OR*

(2) The highest point must not be more than 18.3 meters (60 feet) above the ground.

(d) If your CB station is located near an airport, and if your antenna structure is more than 6.1 meters (20 feet) high, you may have to obey additional restrictions. The highest point of your antenna must not exceed one meter above the airport elevation for every hundred meters of distance from the nearest point of the nearest airport runway. Differences in ground elevation between your antenna and the airport runway may complicate this formula. If your CB station is near an airport, you may contact the nearest FCC field office for a worksheet to help you figure the maximum allowable height for your antenna. Consult Part 17 of the FCC's Rules for more information.

WARNING: Installation and removal of CB station Antennas near powerlines is dangerous. For your safety follow the installation directions included with your antenna.

§95.409 (CB Rule 9) What equipment may I use at my CB station?

(a) You must use an FCC type-accepted CB transmitter at your CB station. You can identify an FCC type-accepted transmitter by the type acceptance label placed on it by the manufacturer. You may examine a list of type-accepted equipment at any FCC Field Office or at FCC Headquarters. Use of a transmitter which is not FCC type-accepted voids your authority to operate the station.

(b) You must not make, or have made, any internal modification to a type-accepted CB transmitter. (See CB Rule 25, §95.425). Any internal modification to a type-accepted CB transmitter cancels the type-acceptance, and use of such a transmitter voids your authority to operate the station.

§95.410 (CB Rule 10) How much power may I use?

(a) Your CB station transmitter power output must not exceed the following values under any conditions:

AM (A3) — 4 watts (carrier power)

SSB — 12 watts (peak envelope power)

(b) If you need more information about the power rule, see the technical rules in Subpart E of Part 95.

(c) Use of a transmitter which has carrier or peak envelope power in excess of that authorized voids your authority to operate the station.

§95.411 (CB Rule 11) May I use power amplifiers?

(a) You may not attach the following items (power amplifiers) to your type-accepted CB transmitter in any way:

(1) External radio frequency (RF) power amplifiers (sometimes called linears or linear amplifiers): *OR*

(2) Any other devices which, when used with a radio transmitter as a signal source, are capable of amplifying the signal.

(b) There are no exceptions to this rule and use of a power amplifier voids your authority to operate the station.

(c) The FCC will presume you have used a linear or other external RF power amplifier if —

(1) It is in your possession or on your premises; *AND*

(2) There is other evidence that you have operated your CB station with more power than allowed by CB Rule 10, §95.410.

(d) Paragraph (c) of this section does not apply if you hold a license in another radio service which allows you to operate an external RF power amplifier.

§95.412 (CB Rule 12) What communications may be transmitted?

(a) You may use your CB station to transmit two-way plain language communications. Two-way plain language communications are communications without codes or coded messages. Operating signals such as "ten codes" are not considered codes or coded messages. You may transmit two-way plain language communications only to other CB stations, to units of your own CB station or to authorized government stations on CB frequencies about —

(1) Your personal or business activities or those of members of your immediate family living in your household.

(2) Emergencies (see CB Rule 18, §95.418);

(3) Traveler assistance (see CB Rule 18, §95.418); *OR*

(4) Civil defense activities in connection with official tests or drills conducted by, or actual emergencies announced by, the civil defense agency with authority over the area in which your station is located.

(b) You may use your CB station to transmit a tone signal only when the signal is used to make contact or to continue communications. (Examples of circuits using these signals are tone operated squelch and selective calling circuits.) If the signal is an audible tone, it must last no longer than 15 seconds at one time. If the signal is a subaudible tone, it may be transmitted continuously only as long as you are talking.

(c) You may use your CB station to transmit one-way communications (messages which are not intended to establish communications between two or more particular CB stations) only for emergency communications, traveler assistance, brief tests (radio checks) or voice paging.

§95.413 (CB Rule 13) What communications are prohibited?

(a) You must not use a CB station —

(1) In connection with any activity which is against federal, state or local law;

(2) To transmit obscene, indecent or profane words, language or meaning;

(3) To interfere intentionally with the communications of another CB station;

(4) To transmit one-way communications, except for emergency communications, traveler assistance, brief tests (radio checks), or voice paging;

(5) To advertise or solicit the sale of any goods or services;

(6) To transmit music, whistling, sound effects or any material to amuse or entertain;

(7) To transmit any sound effect solely to attract attention;

(8) To transmit the word "MAYDAY" or any other international distress signal, except when your station is located in a ship, aircraft or other vehicle which is threatened by grave and imminent danger and you are requesting immediate assistance;

(9) To communicate with, or attempt to communicate with any CB station more than 250 kilometers (155.3 miles) away;

(10) To advertise a political candidate or political campaign; (you may use your CB radio for the business or organizational aspects of a campaign, if you follow all other applicable rules);

(11) To communicate with stations in other countries, except General Radio Service stations in Canada; *OR*

(12) To transmit a false or deceptive communication.

(b) You must not use a CB station to transmit communications for live or delayed rebroadcast on a radio or television broadcast station. You may use your CB station to gather news items or to prepare programs.

§95.414 (CB Rule 14) May I be paid to use my CB station?

(a) You may not accept direct or indirect payment for transmitting with a CB station.

(b) You may use a CB station to help you provide a service, and be paid for that service, as long as you are paid only for the service and not for the actual use of the CB station.

§95.415 (CB Rule 15) Who is responsible for communications I make?

You are responsible for all communications which are made by you from a CB station.

§95.416 (CB Rule 16) Do I have to limit the length of my communications?

(a) You must limit your CB communications to the minimum practical time.

(b) If you are communicating with another CB station or stations, you, and the stations communicating with you, must limit each of your conversations to no more than five continuous minutes.

(c) At the end of your conversation, you, and the stations communicating with you, must not transmit again for at least one minute.

§95.417 (CB Rule 17) Do I identify my CB communications?

(a) You need not identify your CB communications.

(b) [You are encouraged to identify your CB communications by any of the following means:

(1) Previously assigned CB call sign;

(2) K prefix followed by operator initials and residence zip code;

(3) Name; *OR*

(4) Organizational description including name and any applicable operator unit number.]

(c) [You are encouraged to use your "handle" only in conjunction with the methods of identification listed in paragraph (b) of this section.]

§95.418 (CB Rule 18) How do I use my CB station in an emergency or to assist a traveler?

(a) You must at all times and on all channels, give priority to emergency communications.

(b) When you are directly participating in emergency communications, you do not have to comply with the rule about the length of transmissions (CB Rule 16, §95.416). You must obey all other rules.

(c) You may use your CB station for communications necessary to assist a traveler to reach a destination or to receive necessary services. When you are using your CB station to assist a traveler, you do not have to obey the rule about length of transmissions (CB Rule 16, §95.416). You must obey all other rules.

§95.419 (CB Rule 19) May I operate my CB station transmitter by remote control?

(a) You may not operate a CB station transmitter by radio remote control.

(b) You may operate a CB transmitter by wireline remote control if you obtain specific approval in writing from the FCC. To obtain FCC approval, you must show why you need to operate your station by wireline remote control. Send your request and justification to FCC, Gettysburg, Pa 17325. If you receive FCC approval, you must keep the approval as part of your station records. (See CB Rule 27, §95.427.)

(c) Remote control means operation of a CB transmitter from any place other than the location of the CB transmitter. Direct mechanical control or direct electrical control by wire from some point on the same premises, craft or vehicle as the CB transmitter is not considered remote control.

§95.420 (CB Rule 20) May I connect my CB transmitter to a telephone?

(a) You may connect your CB station transmitter to a telephone if you comply with all of the following:

(1) You or someone else must be present at your CB station and must —

(i) Manually make the connection (the connection must not be made by remote control);

(ii) Supervise the operation of the transmitter during the connection;

(iii) Listen to each communication during the connection; *AND*

(iv) Stop all communications if there are operations in violation of these rules.

(2) Each communication during the telephone connection must comply with all of these rules.

(3) You must obey any restriction that the telephone company places on the connection of a CB transmitter to a telephone.

(b) The CB transmitter you connect to a telephone must not be shared with any other CB station.

(c) If you connect your CB transmitter to a telephone, you must use a phone patch device which has been registered with the FCC.

Other Things You Need To Know

§95.421 (CB Rule 21) What are the penalties for violating these rules?

(a) If the FCC finds that you have willfully or repeatedly violated the Communications Act or the FCC Rules, you may have to pay as much as $2,000 for each violation, up to a total of $5,000. (See Section 503(b) of the Communications Act.)

(b) If the FCC finds that you have violated any section of the Communications Act or the FCC Rules, you may be ordered to stop whatever action caused the violation. (See Section 312(b) of the Communications Act.)

(c) If a federal court finds that you have willfully and knowingly violated any FCC Rule, you may be fined up to $500 for each day you committed the violation. (See Section 502 of the Communications Act.)

(d) If a federal court finds that you have willfully and knowingly violated any provision of the Communications Act, you may be fined up to $10,000 or you may be imprisoned for one year, or both. (See Section 501 of the Communications Act.)

§95.422 (CB Rule 22) How do I answer correspondence from the FCC?

(a) If it appears to the FCC that you have violated the Communications Act or these rules, the FCC may send you a discrepancy notice.

(b) Within the time period stated in the notice, you must answer with:

(1) A complete written statement about the apparent discrepancy;

(2) A complete written statement about any action you have taken to correct the apparent violation and to prevent it from happening again: *AND*

(3) The name of the person operating at the time of the apparent violation.

(c) If the FCC sends you a letter asking you questions about your CB radio station or its operation, you must answer each of the questions with a complete written statement within the time period stated in the letter.

(d) You must not shorten your answer by references to other communications or notices.

(e) You must send your answer to the FCC office which sent you the notice.

(f) You must keep a copy of your answer in your station records. (See CB Rule 27, §95.427.)

§95.423 (CB Rule 23) What must I do if the FCC tells me that my CB station is causing interference?

(a) If the FCC tells you that your CB station is causing interference for technical reasons you must follow all instructions in the official FCC notice. (This notice may require you to have technical adjustments made to your equipment.)

(b) You must comply with any restricted hours of CB station operation which may be included in the official notice.

§95.424 (CB Rule 24) How do I have my CB station transmitter serviced?

(a) You may adjust an antenna to your CB transmitter and you may make radio checks. (A radio check means a one way transmission for a short time in order to test the transmitter.)

(b) Each internal repair and each internal adjustment to your FCC type-accepted CB transmitter (See CB Rule 9, §95.409) must be made by or under the direct supervision of a person licensed by the FCC as a General Radiotelephone Operator.

(c) Except as provided in paragraph (d) of this section, each internal repair and each internal adjustment of a CB transmitter in which signals are transmitted must be made using a nonradiating ("dummy") antenna.

(d) Brief test signals (signals not longer than one minute during any five minute period) using a radiating antenna may be transmitted in order to:

(1) Adjust an antenna to a transmitter;

(2) Detect or measure radiation of energy other than the intended signal; *OR*

(3) Tune a receiver to your CB transmitter.

§95.425 (CB Rule 25) May I make any changes to my CB station transmitter?

(a) You must not make or have anyone else make any internal modification to your CB transmitter.

(b) Internal modification does not include:

(1) Repair or servicing of a CB station transmitter (see CB Rule 24, §95.424);

OR

(2) Changing plug-in modules which were type-accepted as part of your CB transmitter.

(c) You must not operate a CB transmitter which has been modified by anyone in any way, including modification to operate on unauthorized frequencies or with illegal power. (See CB Rules 9 and 11, §95.409 and 95.411.)

§95.426 (CB Rule 26) Do I have to make my CB station available for inspection?

(a) If an authorized FCC representative requests to inspect your CB station, you must make your CB station and records available for inspection.

(b) A CB station includes all of the radio equipment you use.

§95.427 (CB Rule 27) What are my station records?

Your station records include the following documents, as applicable.

(a) A copy of each response to an FCC violation notice or an FCC letter. (See CB Rule 22, §95.422.)

(b) Each written permission received from the FCC. (See CB Rule 19, §95.419.)

§95.428 (CB Rule 28) How do I contact the FCC?

(a) Write to your nearest FCC Field Office if you:

(1) Want to report an interference complaint; *OR*

(2) Want to know if the FCC has type-accepted a transmitter for CB.

This whimsical ID card from the "Federal Conmunications Conmission" is a parody of the card carried by an actual FCC field agent. It was made up by a CB'er to pass out to his on-the-air friends.

FEDRAL CONMUNICATIONS CONMISSION
WAHSINGTON
NO. 223-7

This is to certify that

____

is not an officer of the Fedral Conmunications Conmission Field Oporations Section and is not charged with the duty of inspecting, during any reasonable hour, licensed radio stations, investigating violations of radio law, treaties and regulations, and the performance of such other duties as are required. His absence of authority will be respected accordingly by all concerned.

INDEX

19

Pet Peeves

Most of the time, CB is useful and great. Still, after having been an active on the channels since 1959, I do have some pet peeves, gripes, and a couple of sour grapes. I'll admit, they do change from time to time, and vary in intensity. So, if you'll permit a static-numb participant a few paragraphs, I'll share them with you.

1. 10-36's: People come on the air and ask for a 10-36-- that means that they want you to tell them the correct time. This is super dumb. Is it that they can't tell time, or have they spent so much money on their CB equipment that they couldn't afford a wristwatch? They give watches away free with breakfast cereal and cigarettes-- even fast food places sometimes give them away with the greaseburgers. These are nice watches with large LCD numbers to make it easy for those who have trouble figuring out what it means when the big hand is here and the little hand is there.

Let's all fight back at 10-36's by giving the guy back a time check that is about 18-minutes out of kilter. Chances are that the dork who asked for the 10-36 will come right back and tell you that you gave him the wrong time. That's your cue to ask him why he wanted you to tell him the time in the first place if he knew it all along. An alternate approach-- if you can't beat 'em, join 'em. When someone asks for and receives a 10-36, wait about 5 seconds and ask for one yourself-- then have everybody else in your car or hanging around the CB shack get on the air and take turns every 5 seconds asking for additional time checks.

2. Radio Checks: Why do some people seem to ask for a dozen radio checks every day of the week? Is their equipment so shoddy or their installation so inept that they live in constant fear of complete system failure and must continually reassure themselves that is in top ready-status just in case of a sudden weather emergency or enemy attack? Why ask for any radio checks

In CB mythology, the "Smokey" is a hideous ogre or troll that roams the nation's highways performing various acts of indignity upon innocent motorists. Although a CB machine serves some small purpose in avoiding contact with this being, inevitably, the "Smokey" is virtually impossible to evade completely.

at all unless there is some specific reason to suspect that the transmitter is malfunctioning? The only response at all I ever give to requests for radio checks in to tell the guy that I just checked my radio and it's still there. Once someone responded to that line by saying, "OK, but I want my radio checked." My reply-- "I think it's still there, too." He seemed quite satisfied.

3. Smokey Paranoia: Playing "Spot the Smokey" and revealing where he is has a certain amount of usefulness, moreover it's an integral part of CB radio. People expect CB'ers to give smokey breaks. Problem is that, after hearing them since about 1974, they've been so overdone that they've become boring-- and, after all, CB really is entertaining as well as useful, isn't it? I already own a radar detector, so I know where Smokey is hiding, so, for me, they serve no practical purpose. What with most smokeys on the highways having ears anyway, the whole thing has lost much of its sneakiness and folksy humor-- at least it doesn't sound as funny as Burt Reynolds did in "Smokey and The Bandit." I've heard smokey reports pointing out police cars parked double by the side of the road with the cops chatting or having lunch. I've heard smokey breaks when the traffic is so heavy that it's moving at 25 m.p.h.-- is someone worried about getting a speeding ticket? I've heard smokey reports saying he's taking pictures eastbound when he's either taking them westbound or not at all. I suppose it's all part of the great challenge of the highways as expressed so aptly in verse:

Smokey in the bushes,
Smokey in the trees,
But Smokey can't catch us,
'Cause we have CB's!

Any time someone asks me if it's "clean and green," I like to toss a bit of paranoia into them by reporting that it's gray and grizzly-- and that I'm riding bumper-to-bumper in what looks to be a Tijuana Taxi road rallye-- that they've got numerous picture takers set up on both sides of the road with Ferrari chase cars giving out greenstamps as if they were trying to make it into the Guinness Book of records. Of course, I never specify where this nightmare is located. Sometimes, if asked how it looks over my shoulder-- I tell 'em that it looks like I've got dandruff.

4. Home Channels: Some channels are staked out by residents of certain localities or members of local CB clubs, the owners of these channels becoming entirely bent out of shape to the point of violence upon any occasion of its attempted use by strangers or members of other clubs. If you find out that such a "closed" club channel exists in your area, go right ahead and use it. Don't get rattled when they read you the riot act for violating their privacy-- in fact, tell them you're sorry! Say you didn't mean to invade their channel but you didn't know any better since you were just released after being held for 9 years in a Iraqi jail on trumped-up spy charges-- you're just driving through town on the way home to see your wife and kids, one of whom needs a kidney transplant. Fact is, you've made it through for those 9 long years supported by your fond memories of CB radio and how you were going to get a chance again to shout that old Channel 6 (or whatever) gang in Elm City (or wherever)-- and what ever became of those guys from 10 years ago? If that doesn't get you a personal apology from everybody on the channel, plus at least two dinner invitations, then you didn't say it right. For at least a couple of weeks, the club that owns the channel will be a bit more careful when an outsider shows up uninvited.

5. Benedictions/Blessings/Etc.: If, like many, you are annoyed by the meaningless push-button repitition of tired old CB catch phrases and slogans at the end of contacts-- don't just sit there, fight back! I mean, do you really believe that the "Good Lord" is going to forget about all of the rotten things you've done in your life and suddenly "take a liking to you" just because some clown on Channel 19 suggested that possibility? Ridiculous, especially since they monitor Channel 9 in Heaven, anyway!

What about suggestions to keep the "shiny side up and the greasy side down"? Do they think that had they not given you this sage advice that you would have gone ahead with arrangements to drive with the greasy side in the "up" mode?

And so it goes with having a good day today and a better day tomorrow, or having yourself one fine day, or being told that they'll catch you on the backstroke or the flip-flop, etc., etc. What's worse, you're sort of expected to be delighted to have such nonsense beamed at you and it's presumed that you'll reciprocate.

Hold back your desire to say "arrrggggggh!!!" Instead, try something along the lines of, "Yesterday some CB'er told me that I'd have a better day today and this morning my old lady took my Mercedes and ran away with the mailman-- what have you got in mind for tomorrow?"

An all-purpose reply: "Your sign-off was dull; your brain is on null!"

6. Advanced Chronic CB-aholism: Some poor souls somehow have totally wigged out on CB radio. Hooking up with them via radio or eyeball can be a real experience-- the phony cornepone twang in the voice, copious use of 10-codes, an overworked CB lingo vocabulary (note overuse of the following: Roger-D, Ayee 10-4, 4-10, A Big 4-10, fer sure, definitely, mercy). These guys all sound alike-- "Mercy, fer sure, we definitely we be doing an eastbound trip on this Four-Two-Eight superslab, hangin' on the back of a bear in a plain green wrapper! So we catch yer on the backstroke and this be that one **Bodacious Barry** goin' 10-7 and 10-10! Izzat a 10-4? C'mon, c'mon! Bye Bye, we gone!"

If such a barrage of gibberish directed at you hasn't made you lose control of the vehicle (from laughter or rage), then give him a taste of his own medicine-- I look upon it as a necessary session in instant stress-relief. Look, there's really little you could ever hope to say that can equal some of the things that CB-aholics can dream up, but you can at least throw back at him enough CB-type doubletalk to confuse the hell out of him and think that maybe he hasn't fully mastered CB-talk. How 'bout a short burst of something like, "Ayee 10-4, driver, fer sure that was a tobias shout! Better do a quickie 10-122 or you're heading for a 10-0 on that old slidewhistle. Izzat a 4-10 or a negator?"

That message will have absolutely no meaning whatsoever, but it will sound very impressive. It will probably silence him for the next hour or so while he tries to decipher what in the hell you told him. OK, it's only an hour of silence-- think of it as a true service to humanity.

7. Technical Mavens & Experts: Some folks derive 99% of their jollies by telling all other CB'ers how absolutely lousy they sound on the air because of several equipment malfunctions which must be criticized at great length. Commonly cited disasters include: loose connections on mike or antenna, off frequency, weak signal, over or undermodulation, scratchy audio, high SWR, splash-- or combinations of those things. Of a more generalized nature, they simply ask every operator contacted for their CB's description and then reply that the set is basically a piece of junk that would work badly no matter what efforts were made to correct the problem(s).

Sure, CB rigs do develop problems. The time to start worrying is when several operators begin mentioning the same problem to you. But one single guy doing a lengthy oration about your miserable signal, especially after you just heard him tell someone else the same thing, is hardly cause for alarm. These people have at their command a smattering of impressive sounding mumbo-jumbo terms they've memorized to intimidate others and imply that their uninvited analysis is vital to your operation. They scare the dickens out of every operator they contact.

If you get suckered into a conversation with one of these leeches, why not go back to him and say that you're sorry but you missed most of what he said because either, A) his modulation kept cutting back and forth between crystal clear and very muffled, or B) there's a loud hum on his carrier that sounds like an open ground connection somewhere, or C) his signal was so weak that a skip station in Chile was walking all over him. No matter how many times he attempts to repeat his doomsay message about your station, stick by your story about his signal. Maybe he'll dismantle his station to find the problem and you'll be off the hook for a week or so.

8. Channel 9 Antics: Channel 9 abuse is rampant. Every time the band opens for skip between the Caribbean and the northeastern states, it becomes a jumble of Spanish speaking stations. Mostly it's people in the two areas attempting to make contact their families without paying for a long-distance telephone call-- they've settled on Channel 9 because it's usually avoided by most operators for local ratchet jawing. Even some Channel 9 emergency monitor teams abuse the channel, tying it up with unnecessary roll calls and practice drills, or handling non-emergency projects (such as parade control) on the frequency. Some of these teams , sadly, have come to the conclusion that they have a franchise of sorts on responding to all Channel 9 calls in their local area, plus a mandate to chase everybody else off of the channel. The few sour apples in the Channel 9 monitoring barrel have always diminished the ability of the many responsible monitoring groups to be accepted seriously by the CB community at large.

Those things off my chest, I now can go 10-7 and QRT. It's been bodacious-- but for now, bye bye, we down, we gone, we catch ya' on the bounceback. A good day today and a better day tomorrow. We be laying all of those good numbers upon ya'-- keep your nose between the ditches and Smokey out of your britches!

Sorry-- couldn't help it! Forgive me!

Break-Break!!
by KK
BOY, ARE YOU A RADIO OPERATOR, MISTER ?
YUP
BOYS, HE'S ONE OF THE TOP CB OPERATORS ON THE BAND !!
GEE
WOW
GEE, THAT'S WHAT I'M GOING TO BE WHEN I GROW UP.....
FAT

INDEX

10-36

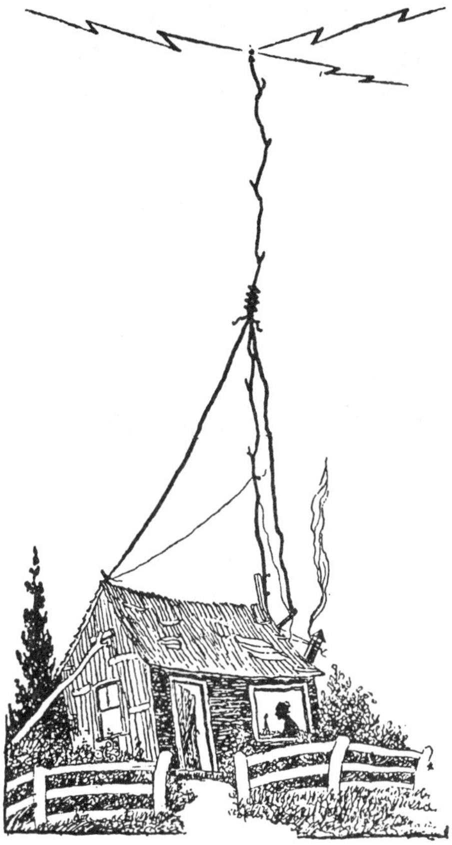

VIRGINIA STATE CITIZENS BAND
RADIO ASSOCIATION
ENGLISH INTERNATIONAL D.X. CLUB
ENGLISH INTERNATIONAL DX··CLUB
HAMPSTEAD C B RADIO CLUB
COMMUNICATIONS VIA SINGLE SIDE BAND
SSB
11 METERS
FIVE
LAKE COUNTY
WATTERS
LAKELAND
C 9 B
INDIANA
CITIZENS COMMUNICATIONS CLUB
CB RADIO
OF SAVANNAH, GEORGIA
QUEEN CITY
C. B. 5 CLUB
WATTERS
CUMBERLAND, MARYLAND
CLUB INTERNATIONAL DX
BRAVO-WHISKY
MONMOUTH
CB
COUNTY
EMERGENCY AID NETWORK
MEMBER
C B
CITIZENS BAND RADIO CLUB
RHEIN MAIN SIDEBANDERS OF GERMANY
ΣΥΛΛΟΓΟΣ ΡΑΔΙΟΣΥΧΝΟΤΗΤΟΣ ΠΟΛΙΤΩΝ ΕΛΛΑΔΟΣ
27
C.B. CLUB OF GREECE
ALFA BRAVO DX-GROUP
CITIZEN'S RADIO
CLUB
OF
WICHITA, INC.
BRAVO NOVEMBER Po Box 6 1
CLUB DX DE BELGIQUE
TREE TOP TALKERS
73's GERMANY 88's
VRA
Z.H.C.B.
27 Mc.
C 11 B
METER
CLUB